STRUCTURE AND BONDING

Volume 36

Editors:
J. D. Dunitz, Zürich · J. B. Goodenough, Oxford
P. Hemmerich, Konstanz · J. A. Ibers, Evanston
C. K. Jørgensen, Genève · J. B. Neilands, Berkeley
D. Reinen, Marburg · R. J. P. Williams, Oxford

With 69 Figures and 22 Tables

Springer-Verlag
Berlin Heidelberg GmbH 1979

ISBN 978-3-662-15418-2 ISBN 978-3-540-35387-4 (eBook)
DOI 10.1007/978-3-540-35387-4

Library of Congress Catalog Card Number 67-11280

Originally published by Springer-Verlag Berlin Heidelberg New York in 1979
Softcover reprint of the hardcover 1st edition 1979

Typesetting: R. & J. Blank, München.
2152/3140-543210

Contents

Manuscripts will be accepted by the editors:

Professor Dr. *Jack D. Dunitz*	Laboratorium für Organische Chemie der Eidgenössischen Hochschule Universitätsstraße 6/8, CH-8006 Zürich
Professor *John B. Goodenough*	Inorganic Chemistry Laboratory University of Oxford, South Parks Road Oxford OX1 3QR, Great Britain
Professor Dr. *Peter Hemmerich*	Universität Konstanz, Fachbereich Biologie Postfach 733, D-7750 Konstanz
Professor *James A. Ibers*	Department of Chemistry, Northwestern University Evanston, Illinois 60201, U.S.A.
Professor Dr. *C. Klixbüll Jørgensen*	Dépt. de Chimie Minérale de l'Université 30 quai Ernest Ansermet, CH-1211 Genève 4
Professor *Joe B. Neilands*	Biochemistry Department, University of California Berkeley, California 94720, U.S.A.
Professor Dr. *Dirk Reinen*	Fachbereich Chemie der Universität Marburg Gutenbergstraße 18, D-3550 Marburg
Professor *Robert Joseph P. Williams*	Wadham College, Inorganic Chemistry Laboratory Oxford OX1 3QR, Great Britain

---

SPRINGER-VERLAG

D-6900 Heidelberg 1
P. O Box 105280
Telephone (06221) 487·1
Telex 04-61723

D-1000 Berlin 33
Heidelberger Platz 3
Telephone (030) 822001
Telex 01-83319

SPRINGER-VERLAG
NEW YORK INC.

175, Fifth Avenue
New York, N.Y. 10010
Telephone (212)477-8200

The Resonance Raman Effect – Review of the Theory and of Applications in Inorganic Chemistry

Robin J.H. Clark and Brian Stewart*

Christopher Ingold Laboratories, University College London, 20 Gordon Street, London WC1H OAJ, England

Table of Contents

* Present Address: Department of Chemistry, University of Glasgow, Glasgow 912 8QQ.

1 Introduction

Raman spectroscopy provided most of the early experimental data on molecular vibrational frequencies, but with the advent of commercial infrared spectrometers in the 1940's, the technique stagnated with respect to infrared spectroscopy. This was despite the fact that Raman spectroscopy enjoyed, and still enjoys, certain natural advantages over infrared spectroscopy, *viz*.

(i) The possibility of working with glass or silica cells, which are much easier to handle and which are less fragile and moisture-sensitive than the alkali-halide windows used in infrared spectroscopy;

(ii) Water is an excellent solvent for Raman studies, but a difficult one for infrared work; this is because the Raman scattering from water is weak whereas the infrared absorption by it is strong;

(iii) For solids, no mulling agents are necessary and so no complications arise from bands of the mulling agent;

(iv) Totally symmetric modes of vibration may be identified by Raman spectroscopy by polarisation studies, even for freely rotating fluids; and

(v) Raman shifts of low frequency can be measured as easily at (say) 100 cm^{-1} as at 1 000 cm^{-1}, whereas the degree of sophistication of infrared equipment must increase rapidly for low frequency studies; this is principally because of the very low intensity of the infrared emission by any source in the low frequency region.

The renaissance of Raman spectroscopy came about in the 1960's by way of several striking technical advances. The most dramatic of these was the discovery of the laser, which emits a coherent beam of light, and which has many advantages over the traditional sources, such as the mercury arc, whose commonly used excitation lines have wavelengths of 435.8 and 546.1 nm. These advantages include the following:

(i) The power flux density of laser beams is much greater than that of traditional sources, e.g. Hg lamp $\sim$ 1 W cm^{-2}, continuous wave (c.w.) Ar^+ ion laser $\sim 10^5$ W cm^{-2}, pulsed glass lasers $\sim 10^{12}$ W cm^{-2}. As the intensity of the scattered radiation is directly related to that of the incident beam, it is clear that much more intense Raman spectra can be obtained by use of laser beams.

(ii) With Ar^+ and Kr^+ ion lasers, there is the choice of about 20 different exciting lines with wavelength ranging from 333.6 to 799.3 nm. Thus the wavelength of the exciting line can be chosen so as either to avoid or alternatively to approach at will the maxima of absorption bands of the scattering molecules.

(iii) Laser beams are very narrow (e.g. for the He/Ne laser, $\Delta\nu_{1/2} = 0.05$ cm^{-1}) and can be made much narrower still by introduction of etalons. High resolution work, e.g. on the vibrational-rotational spectra of molecules in the gas phase, thus becomes more feasible with lasers than with conventional sources.

(iv) The directional properties and linear polarisation of the laser beam make it invaluable for studies of single crystals in which the magnitudes of the individual components, $\boldsymbol{\alpha}_{\rho\sigma}$, of the polarisability tensor governing the intensity of the Raman scattering, can be measured. The directional properties are also an advantage in the study of surfaces of opaque solids and of reaction products formed on the surfaces of electrodes.

The laser was not, however, the only technical advance in the field of Raman spectroscopy to take place in the 1960's. The introduction of extremely sensitive phototubes enabled Raman spectra to be recorded photoelectrically rather than photographically, with consequent enormous saving in recording times. Further, high quality double and triple monochromators of high efficiency have been developed so that it is now possible to scan to within about 3 cm^{-1} of the exciting line and to obtain good quality Raman spectra. The consequence of these advances is that Raman spectroscopy has now reasserted itself so that it has become again the equal partner with infrared spectroscopy in the study of molecular vibrations.

Raman spectra known as resonance Raman spectra are obtained when a molecule in the vapour state at high pressure or in a condensed state is excited with a laser beam whose frequency corresponds or closely corresponds with that of the band maximum of an electric-dipole-allowed transition of the molecule. Such spectra are frequently characterised by an enormous enhancement to the intensity of the band arising from a totally symmetric fundamental of the molecule, together with the appearance of high intensity overtone progressions in this same fundamental. The technique provides detailed information about the chromophore, because it is only vibrational modes closely associated with atoms at the absorbing centre in the molecule which display the effect. The intensification of the Raman band is so great that compounds at concentrations of 10^{-6} M in water can now be detected. The technique has thus attracted considerable attention in biological circles, and much information on the chromophoric centres of molecules such as cytochrome c and haemoglobin has been obtained. The technique is also being used for the remote sensing of small amounts of atmospheric pollutants such as NO_2 and O_3.

Many of the early studies of the resonance Raman effect, particularly of organic molecules possessing conjugated double bonds, were carried out by *Behringer* and *Brandmüller*, and comprehensive reviews of this area are available (*1, 2*). Comparable studies on inorganic complexes have only been carried out in the last ten years, the results till 1974 having been reviewed previously (*3*). However, within the last few years there has been an enormous increase in activity (theoretical as well as experimental) in this area, and accordingly it seems appropriate to review the subject as it now stands. Particularly important would appear to be the advances in our understanding of the different mechanisms of Raman scattering and of excitation profiles (plots of Raman band intensities versus excitation wavenumber), and the information these give pertinent to the assignment of the resonant electronic transitions.

2 Résumé of the Theory of the Resonance Raman Effect

2.1 The Dispersion Equation

The total intensity of radiation scattered into a solid angle of 4π during a Raman process in which a molecular system, initially in a state $\langle G|$, makes a transition to a final state $|F\rangle$, is given by

$$I_S^{TOTAL} = (2^7 \pi^5/9)\, I_L\, \nu_S^4 \sum_{\rho,\sigma} |(\alpha_{\rho\sigma})_{G\to F}|^2 \qquad (1)$$

I_L is the intensity of the incident radiation of wavenumber ν_L. ν_S is the wavenumber of the scattered radiation: $\nu_S = \nu_L \pm \nu_{GF}$, where the plus sign refers to anti-Stokes and the minus sign to Stokes scattering. ν_{GF} is the wavenumber of the molecular transition $|G\rangle \to |F\rangle$. $(\alpha_{\rho\sigma})_{G\to F}$ is the polarisability or scattering tensor for the process $|G\rangle \to |F\rangle$ in which the polarisations of the incident and scattered radiation are represented by the indices σ and ρ respectively. ($\sigma, \rho = x, y, z$).

In the usual experimental set up, the scattered radiation is collected in some small solid angle around an observation direction at 90° to that of the incident radiation. A schematic diagram of a typical laser Raman set-up involving 90° collection optics is shown in Fig. 1, together with the definition of the depolarisation ratio

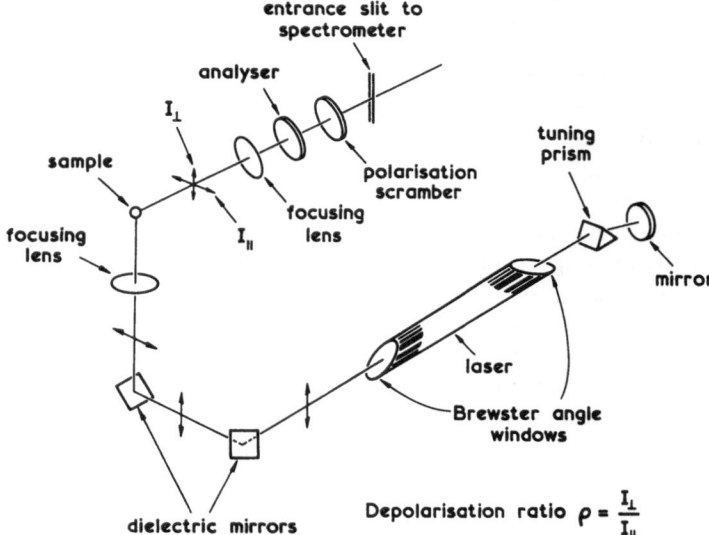

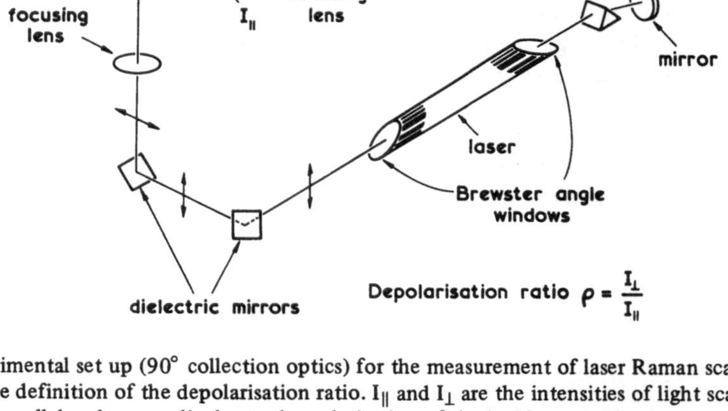

Fig. 1. Experimental set up (90° collection optics) for the measurement of laser Raman scattering, illustrating the definition of the depolarisation ratio. $I_\parallel$ and $I_\perp$ are the intensities of light scattered, respectively, parallel and perpendicular to the polarisation of the incident exciting beam

$\rho_l = I_\perp/I_\parallel$ for linearly polarised incident radiation. In this case, Eq. (1) must be multiplied by the factor $(3/8\,\pi)\,(1 + \rho_l)/(1 + 2\,\rho_l)$. (*5a*). If only the parallel component of scattered radiation is measured the factor is $(3/8\,\pi)\,(1 + 2\,\rho_l)^{-1}$. ($I_\perp = \rho_l\, I_\parallel$). For naturally polarised (unpolarised) or circularly polarised incident light the relevant factors are given in terms of the depolarisation ratio, ρ_n, for natural polarisation, where

$$\rho_n = 2\,\rho_l/(1 + \rho_l)\,.$$

Thus,

$$I(\pi/2) = (3/4\,\pi)\,(2 + \rho_n)^{-1}\, I_S^{TOTAL}$$

and

$$I_\parallel(\pi/2) = (3/8\,\pi)\,[(2 - \rho_n)/(2 + \rho_n)]\, I_S^{TOTAL}$$

The polarisability tensor is given by a second-order perturbation expression known as the *Kramers-Heisenberg* dispersion formula (*4*)

$$(\boldsymbol{\alpha}_{\rho\sigma})_{G\rightarrow F} = \frac{1}{hc}\sum_E \left[\frac{\langle F|\boldsymbol{\mu}_\rho|E\rangle\langle E|\boldsymbol{\mu}_\sigma|G\rangle}{\nu_{GE} - \nu_L + i\Gamma_E} + \frac{\langle F|\boldsymbol{\mu}_\sigma|E\rangle\langle E|\boldsymbol{\mu}_\rho|G\rangle}{\nu_{FE} + \nu_L + i\Gamma_E}\right] \qquad (2)$$

where $\boldsymbol{\mu}_\sigma, \boldsymbol{\mu}_\rho$ are the electric-dipole-moment operators e.g. $\boldsymbol{\mu}_\rho = -\sum_k e(r_k)_\rho$, and $(r_k)_\rho$ is the ρ-th component of the position vector of the kth electron. Γ_E is the damping factor or natural half-width of the state $|E\rangle$. It must be noted that the sum over the intermediate states, $|E\rangle$, excludes the initial and final states $|G\rangle$ and $|F\rangle$.

Equation (2) (without damping factors) was originally developed by *Kramers* and *Heisenberg* by analogy with the classical theory of dispersion. The classical problem was the explanation of the lower velocity of light in materials compared to that in the vacuum and this was described in terms of the interference between scattered and unscattered waves. The quantum mechanical description is based on the model of the electric dipole moments induced in the system by the radiation field and the subsequent radiation by these dipoles. The dispersion formula was later verified by *Dirac* (*6*) using quantum electrodynamical arguments. Following *Heitler* (*7*), *Dirac* (*8*) introduced the damping terms in order to prevent the expression becoming infinite in the case of resonance scattering, for which the wavenumber of the incident radiation approaches and becomes equal to that of a molecular transition: $\nu_L = \nu_{GE}$. In these circumstances, the first term in square brackets of Eq. (2) may give rise to a large increase in scattered intensity, this increase being limited by the magnitudes of the appropriate matrix elements in the numerator and of the half-width, Γ_E, for the resonant state $|E\rangle$. The second term in the square brackets of Eq. (2) is non-resonant and produces a slowly varying background contribution, which is usually negligible compared to the resonant contribution when ν_L is in the resonance region. However, both halves of the polarisability expression must be taken into account when ν_L is far from resonance, as in the case in the normal Raman effect for which $\nu_{GE} \gg \nu_L \gg \nu_{GF}$.

This review is largely concerned with vibrational Raman scattering in which the states $|G\rangle$ and $|F\rangle$ differ only in the possession of one or more quanta of vibrational energy. In systems with electronically degenerate ground states it is impossible to avoid entirely the effects of simultaneous changes in the electronic and vibrational quantum numbers: This latter point is discussed briefly later on with respect to the Jahn-Teller effect as well as the electronic Raman effect.

2.2 *The Vibronic Nature of the Scattering Tensor*

To proceed further it is necessary to elaborate on the vibronic nature of the states and operators in the scattering tensor. The adiabatic Born-Oppenheimer (ABO) approximation will be employed in which the vibronic states are formed by products of pure vibrational states $\Lambda_{iu}(Q)$ with pure electronic states $\Phi_i(Q, r)$. The latter are referred to some fixed positions of the nuclei and are therefore parametrically dependent on the internuclear distances expressed as a set, Q_ξ, of normal coordinates. r is a set of electronic coordinates, and i, u are electronic and vibrational quantum numbers respectively. Thus

$$\Psi_{i,u}^{ABO}(Q_\xi, r) = \Phi_i(Q_\xi, r)\, \Lambda_{i,u}(Q_\xi) \tag{3}$$

and

$$\Lambda_{i,u}(Q_\xi) = \prod_\xi \phi_{i,u}^\xi(Q_\xi)$$

where $\phi_{i,u}^\xi$ is usually a harmonic oscillator function for the ξ-th normal coordinate, Q_ξ, in the electronic state $|i\rangle$. In many instances it is necessary to consider only a single normal coordinate and therefore the index ξ will be dropped for these cases. A vibrational level will be conveniently written as $|u_i\rangle$.

The validity of the ABO approximation depends on the separability of the electronic and nuclear vibrational motions, which is contingent upon the smallness of the ratio of the electronic to the nuclear mass. The electronic part of the ABO product function [Eq. (3)] is accordingly defined for fixed nuclear positions and is known as a crude-adiabatic (CBO) electronic function. A CBO electronic function is an eigenfunction of a total electronic Hamiltonian which contains the electron-nuclear and nuclear-nuclear potential energies (for some fixed positions of the nuclei), but lacks the nuclear kinetic-energy terms. This approximation is a good one, provided the electronic states have large energy separations compared with a vibrational quantum, and it has the advantage that an adiabatic potential-energy curve for nuclear vibration is given by a plot of the total electronic energy for various fixed positions of the nuclei. When electronic states approach very closely in energy, strong vibronic coupling leads to narrowly avoided crossings of the adiabatic potential curves and in these regions the (neglected) nuclear momentum may cause non-adiabatic coupling between different ABO manifolds. In the diabatic BO (DBO) approximation, the basis states exactly diagonalise the kinetic energy terms in the Hamiltonian and this gives a better description of the system near the avoided crossings but is a worse approximation everywhere else (*30*).

When the ABO states are substituted into the polarisability expression (2) it is possible formally to complete the integrations over the electronic coordinates in the matrix elements. Thus, for example,

$$\langle G | \boldsymbol{\mu}_\rho | E \rangle = \langle \Lambda_{g,u} | \boldsymbol{\varrho}_{ge}(Q) | \Lambda_{e,v} \rangle \tag{4}$$

where

$$\boldsymbol{\varrho}_{ge}(Q) = \int \Phi_g(Q, r)\, \boldsymbol{\mu}_\rho\, \Phi_e(Q, r)\, dr$$

and $\boldsymbol{\sigma}_{ge}(Q)$ is defined similarly.
This is because the electric dipole operators, $\boldsymbol{\mu}_\rho$, act only on the electronic coordinates, the contribution to the transition moment from the nuclear coordinates being negligible. The resulting matrix elements, $\boldsymbol{\varrho}_{ge}(Q)$, are still parametrically dependent on Q through the electronic states. This dependence is supposed to be a weak function of the internuclear coordinates and is therefore described by a rapidly converging Taylor series expanded about the equilibrium configuration, Q = 0, of the ground electronic state:

$$\boldsymbol{\varrho}(Q) = \boldsymbol{\varrho}(0) + \left(\frac{\partial \boldsymbol{\varrho}(Q)}{\partial Q}\right)_{Q=0} Q + \ldots \tag{5}$$

If this expansion for the transition moments is used then the resonant part of the scattering tensor may be expressed, in terms to the first order in Q, as follows:

$$\begin{aligned}(\boldsymbol{\alpha}_{\rho\sigma})_{u\to w} = \sum_e \sum_v (E_{ev} - E_{gu} - E_L + i\,\Gamma_{ev})^{-1} \\ \times [\ \boldsymbol{\varrho}^0_{ge}\,\boldsymbol{\sigma}^0_{eg}\,\langle w_g | v_e \rangle \langle v_e | u_g \rangle \\ + \boldsymbol{\varrho}'_{ge}\,\boldsymbol{\sigma}^0_{eg}\,\langle w_g | Q | v_e \rangle \langle v_e | u_g \rangle \\ + \boldsymbol{\varrho}^0_{ge}\,\boldsymbol{\sigma}'_{eg}\,\langle w_g | v_e \rangle \langle v_e | Q | u_g \rangle]\end{aligned} \tag{6}$$

using the abbreviations $\boldsymbol{\varrho}^0 = \boldsymbol{\varrho}(0)$, $\boldsymbol{\varrho}' = (\partial \boldsymbol{\varrho}(Q)/\partial Q)_{Q=0}$, etc. E_{ev} is the energy of the v-th vibrational level of the intermediate electronic state $|e\rangle$, E_{gu} the energy of the u-th level of the ground state $|g\rangle$ and $E_L (= hc\,\nu_L)$ the energy of the incident light quantum. In the harmonic approximation, $E_{ev} = E_{e0} + vhc\nu$ for a fundamental of wavenumber ν. The circular frequency $\omega (= 2\pi c\nu)$ is employed by many physicists, who therefore express vibrational energies as $\hbar\omega$.

The first term in the square brackets of Eq. (6) gives rise to the A term and the second and third terms to the B term of *Albrecht* (*9, 10*). The further analysis of this expression requires specification of the energy region of excitation.

2.3 Normal Raman Scattering

In the ideal limit of the normal (non-resonance) Raman effect the energy of the exciting radiation is far removed from any transition energy of the system. That is

$$E_{ev} - E_{gu} \gg E_L \gg \hbar\omega$$

The energy denominators in the resonant and non-resonant parts of the polarisability [Eq. (2)] are each large and insensitive to the vibrational quantum numbers, v_e, of the intermediate electronic state $|e\rangle$. Since the v_e represent a complete orthonormal set of states, the sums over them in Eq. (6) may be evaluated by using the closure theorem, provided that the v-dependence of the energy denominators is negligible. The closure theorem may be stated essentially as: $\sum_{\text{all}\,v} |v\rangle\langle v| \equiv 1$, and arises from the matrix product rule $\sum_j \mathbf{A}_{ij}\,\mathbf{B}_{jk} = (\mathbf{AB})_{ik}$. This leads to the following simplifications in the vibrational matrix elements:

The matrix elements in the non-resonant part of the scattering tensor may be treated in the same manner. The final expression for Eq. (6), plus its non-resonant counterpart, is

$$\left.\begin{aligned} &\sum_v \langle w_g|v_e\rangle\langle v_e|u_g\rangle = \langle w_g|u_g\rangle = \delta_{u,w} \\ &\sum_v \langle w_g|Q|v_e\rangle\langle v_e|u_g\rangle = \sum_v \langle w_g|v_e\rangle\langle v_e|Q|u_g\rangle = \langle w_g|Q|u_g\rangle = \delta_{u,w\pm1} \end{aligned}\right\} \tag{7a}$$

$$\delta_{u,w\pm1} = \left(\frac{\hbar}{2\mu\omega}\right)^{1/2} [w^{1/2}\,\delta_{u,w-1} + (w+1)^{1/2}\,\delta_{u,w+1}] \tag{7b}$$

$\delta_{u,w} = 1$ if $u = w$ and $= 0$ otherwise, as a result of the orthonormality of the vibrational states. $\delta_{u,w\pm1}$ is defined in Eq. (7b) and results from the selection rule on the matrix elements of Q in the harmonic oscillator approximation (*11*).

$$\begin{aligned} (\alpha_{\rho\sigma})_{u\to w} = &\sum_e \left[\frac{\rho^0_{ge}\,\sigma^0_{eg}}{E_e - E_g - E_L} + \frac{\sigma^0_{ge}\,\rho^0_{eg}}{E_e - E_g + E_L}\right]\delta_{u,w} \\ &+ \sum_e \left[\frac{\rho'_{ge}\,\sigma^0_{eg} + \rho^0_{ge}\,\sigma'_{eg}}{E_e - E_g - E_L} + \frac{\sigma'_{ge}\,\rho^0_{eg} + \sigma^0_{ge}\,\rho'_{eg}}{E_e - E_g + E_L}\right]\delta_{u,w\pm1} \end{aligned} \tag{8}$$

The half band-widths have been omitted on the grounds that $E_e - E_g \pm E_L \gg \Gamma_e$ for the case of normal Raman scattering. More concisely, Eq. (8) may be written

$$(\alpha_{\rho\sigma})_{u\to w} = \alpha^0_{\rho\sigma}\,\delta_{u,w} + \alpha'_{\rho\sigma}\,\delta_{u,w\pm1}$$

since

$$\rho'\,\sigma^0 + \rho^0\,\sigma' = (\rho\sigma)'\,.$$

This represents Placzek's results to the first order (*5b*). The first term gives rise only to Rayleigh scattering (u = w) and depends on the electronic polarisability of the scattering species at equilibrium in its ground state. The second term gives a Raman fundamental (u = w ± 1 for anti-Stokes and Stokes scattering, respectively) and depends on the first derivative of the polarisability with respect to the normal coordinate involved. To account for overtones which, however, are generally very weak or absent in the normal Raman effect, higher derivatives of the polarisability must be considered. In general,

$$\boldsymbol{\alpha}_{\rho\sigma} = \sum_{n=0}^{\infty} \frac{1}{n!} \frac{\partial^n(\boldsymbol{\alpha}_{\rho\sigma})}{\partial Q^n}$$

where the even derivatives contribute to Rayleigh scattering and to even harmonics (even multiples of the fundamental). Odd derivatives contribute to the fundamental scattering and the odd harmonics.

In summary, normal Raman scattering occurs via an intermediate state described by comparable contributions from a large number of excited electronic states, e, of the system. As a result the intermediate state has no well-defined electronic symmetry and the scattering tensor is symmetric in the indices ρ and σ. The symmetry of the tensor arises also if the contribution from the two halves of Eq. (2) are equal, which is true to a good approximation when $\nu_L \ll \nu_{GE} \cong \nu_{FE}$. This depends on the fact that, in the sum over E, the higher energy states predominate despite their unfavourable denominators. The states near and above the dissociation limit are important, presumably because of their high density and oscillator strengths, and there is a good empirical proportionality between the polarisability and the first ionisation potential *(12)*. The insensitivity of the scattering to the vibrational structure of the intermediate states accounts for the zero contribution from the vibrational overlap integrals (Franck-Condon factors). This latter point, together with the fact that the second and higher derivatives of $\boldsymbol{\alpha}$ are usually negligibly small, explains the general weakness or absence of overtones in normal Raman scattering.

2.4 Resonance Raman Scattering

As resonance is achieved by allowing the incident light energy, E_L, to approach some molecular transition energy, $E_{ev} - E_{gu}$, the denominator $(E_{ev} - E_{gu} - E_L + i\,\Gamma_{ev}) = \Delta E + i\,\Gamma_{ev}$ becomes small and the scattering tensor correspondingly large. When ΔE is of the order of a vibrational energy-level difference, $\hbar\omega$, then obviously the approximation of neglecting the dependence of the denominators on the excited-state vibrational quantum numbers becomes inadequate. In this case, closure cannot be applied as it was in Eq. (7a) to reduce the polarisability to being a property of the ground state only. However, simplifications also take place in that, as E_L approaches some particular transition energy, the appropriate excited state will dominate the sum over states in the scattering tensor. It is therefore generally sufficient to consider only one, or at most two, electronic manifolds in accounting for resonance Raman scattering.

It is useful to discuss the factors determining the relative importance of the zeroth-order (Franck-Condon, FC) and first-order (Herzberg-Teller, HT) terms in the nuclear-coordinate expansion of the polarisability. In an early treatment of the resonance Raman effect by *Albrecht* 1961 (*9*), it was assumed that the closure procedure was still a good approximation even in regions of excitation close to absorption bands. If this were true then, as we have seen, the FC terms would be responsible only for Rayleigh scattering. The enhancement of Raman scattering must then be attributed to the first-order terms. These were explicitly developed by Albrecht from a (first-order) Herzberg-Teller perturbation description (*13, 14*) of vibronic coupling:

$$\boldsymbol{\varrho}'_{ge} = \sum_{s \neq e} \boldsymbol{\varrho}^0_{gs} \langle s | \partial \mathscr{H} / \partial Q | e \rangle_0 (E_e - E_s)^{-1} \tag{9}$$

where

$$\langle s | \partial \mathscr{H} / \partial Q | e \rangle_0 \quad \text{is written for} \quad \int \Phi_s(Q, r) \left(\frac{\partial \mathscr{H}}{\partial Q}\right)_{Q=0} \Phi_e(Q, r) \, dr$$

and $\mathscr{H}$ is the Hamiltonian for the total electronic energy of the molecule. In this approach, the derivative of the transition moment arises because the variation in the Hamiltonian with respect to the normal coordinate, Q, can mix the state $|e\rangle$ with other states $|s\rangle$ of the appropriate symmetry. Thus, by Eq. (9), a portion of the transition moment, $\boldsymbol{\varrho}^0_{gs}$, is acquired for the process $|g\rangle \rightarrow |e\rangle$. This mechanism is particularly useful when the transition to the state of interest is forbidden in zero order, i.e. $\boldsymbol{\varrho}^0_{ge} = 0$ as, for example, in the Laporte-forbidden d–d or f–f transitions. However, since large scattering intensities are favoured by a strong intrinsic transition moment, $\boldsymbol{\sigma}^0_{eg}$, in the relevant part of Eq. (6):

$$(\boldsymbol{\alpha}_{\rho\sigma})_{u \rightarrow w} = \sum_v (E_{ev} - E_{gu} - E_L + i\,\Gamma_{ev})^{-1} \boldsymbol{\varrho}'_{ge} \boldsymbol{\sigma}^0_{eg} \langle w_g | Q | v_e \rangle \langle v_e | u_g \rangle$$

[with $\boldsymbol{\varrho}'_{ge}$ given by Eq. (9)], this approach led to the statement (9) that the modes enhanced in resonance are those most responsible for "forbidden" intensity in an *allowed* electronic band. This is true in the case of an allowed transition with FC terms which are accidentally very small or zero so that the HT terms then dominate the scattering mechanism. This implies that totally symmetric modes are only enhanced if they are active in vibronic coupling between two allowed electronic bands *of the same symmetry*. However, the situation just outlined is not generally encountered. More usually, allowed transitions have non-zero and often large FC integrals, and resonance scattering is dominated by these because the closure sum rule is not even approximately valid close to, or within an absorption band. Furthermore, a more general treatment must include vibronic coupling within a single electronic excited state via totally symmetric modes. Actually, these last two points are intimately related, since the one-state vibronic coupling (via a totally symmetric mode in a non-degenerate state) can be shown to give rise to the Franck-Condon effect; that is, extended overtone progressions in the frequency of the responsible totally symmetric mode. Progressions in the excited state frequency appear in an absorption spectrum at low temperature (Fig. 2). The ground-state frequency appears in relaxed fluorescence (Fig. 2) and in the resonance Raman effect (Fig. 3). In a converse approach, *Tang* and *Albrecht* (*15*) have shown that, when the nuclear-coordinate dependence of the energy denominators is explicitly taken into account in the A term, the FC contribution becomes equivalent to a one-state "B term". The relation between the vibronic view and the overlap view of FC scattering is discussed in more detail in Section 2.6.

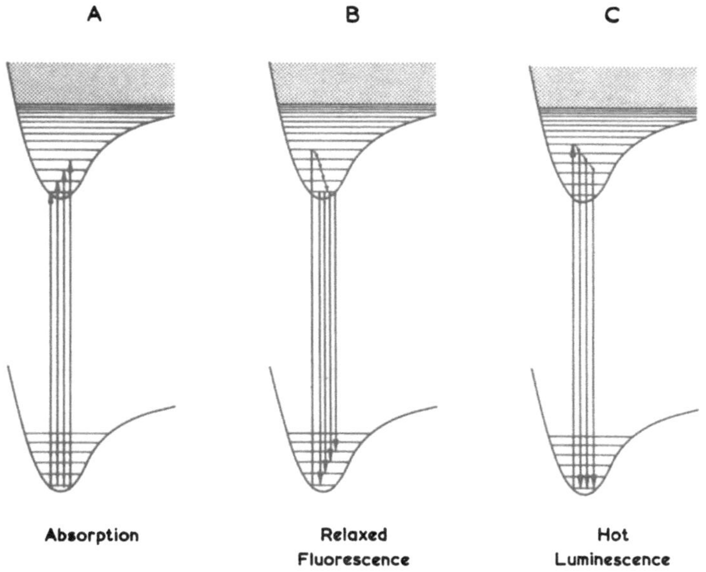

Fig. 2. The origin of vibrational progressions in absorption and emission processes.
A. A progression in the excited-state frequency occurs in absorption. B. A progression in the ground-state frequency occurs in relaxed fluorescence where non-radiative relaxation processes occur faster than the emission process in the excited state. C. Excited state vibrational intervals may occur in hot luminescence where the speed of non-radiative relaxation is similar to, or slower than, that of the emission process

Since the H-T terms are developed from the first-order terms active in the non-resonance Raman effect, they are usually the most important source of preresonance enhancement whereas the FC terms are active in proportion to the extent to which closure fails for the vibrational levels of the intermediate state. Some physical insight into these different scattering mechanisms is furnished by the following interpretation of the energy denominators of the scattering tensor.

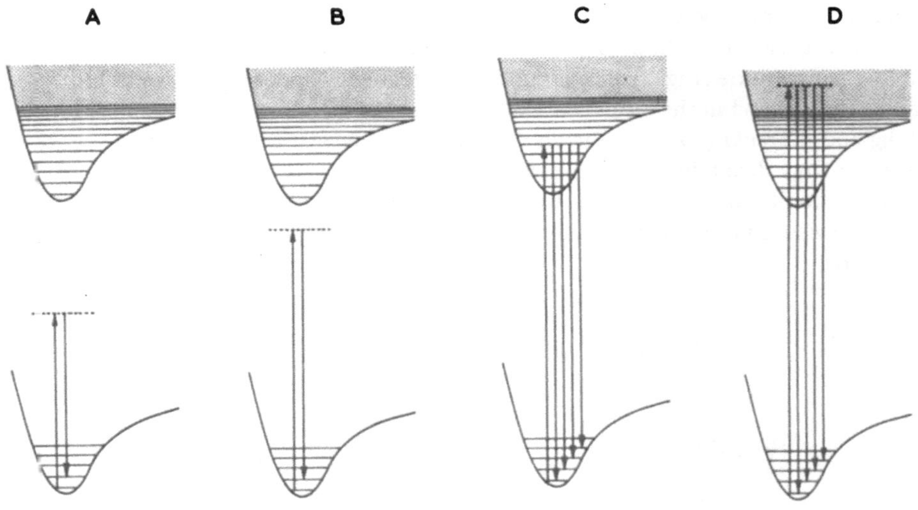

Fig. 3. Types of Raman scattering, depending on the proximity of the exciting light energy to the molecular transition energies.
A. Normal Raman scattering with excitation far from any real molecular transitions. B. Pre-resonance Raman scattering in which the excitation energy approaches that of a molecular transition. Enhancement to the scattering by fundamentals is observed. C. Excitation energy in the region of discrete rotational-vibrational levels of a single electronic intermediate state. This situation has been termed discrete resonance-Raman scattering (*23*) irrespective of whether or not the resonance is exact with a particular discrete level (resonance fluorescence limit). Overtone scattering may occur for totally symmetric modes with displaced potential curves in the excited state. D. The incident light energy is in the range of dissociative continuum levels. Overtone scattering occurs as in C

In the Raman effect, scattering occurs when the molecular system, interacting with the incident radiation, makes a transition from an initial stationary state to an intermediate state from which, by a second interaction, a transition occurs to the final stationary state. The intermediate state is not in general a stationary state of the molecular system. This virtual intermediate state is not the solution of a time-independent Schrödinger equation and therefore does not correspond to a well-defined value of the energy. The intermediate state is expanded in terms of the excited molecular eigenstates. In this expansion each eigenstate may be thought of as weighted in inverse proportion to the extent by which it fails to conserve energy in the intermediate state. This is expressed in the resonant part of the scattering tensor by the mismatch between the energy of the exciting photon and the appropriate formal transition energy: $\Delta E = (E_E - E_G) - E_L$. Now, if a state fails to conserve energy by an amount ΔE, then it is permitted a lifetime, Δt, given by the Heisenberg relation

$$\Delta E \cdot \Delta t \sim \hbar$$

So, when closure over the vibrational levels begins to fail as resonance is approached and $\Delta E \lesssim \hbar\omega$, the implication is that the lifetime of the intermediate state is given by $\Delta t \gtrsim \omega^{-1}$. In the limit of exact resonance, when $\Delta E = 0$, the lifetime is limited by the natural lifetime, $\hbar\Gamma_E^{-1}$, which includes contributions from radiative, collisional and Doppler broadening, etc. Thus, the onset of FC scattering is associated with scattering times of the order of a vibrational period or longer. This is consistent with the picture of an established potential-energy curve for nuclear vibration in the excited state. These ideas are supported by measurements of scattering times using pulsed excitation in diatomic systems (*16*). For exact resonance with a discrete rotational-vibrational transition, it is found that scattering times are typically in the range $10^{-5} - 10^{-8}$ s (cf. $\omega^{-1} \sim 10^{-11} - 10^{-13}$ s). Off resonance, scattering times become very short but, because of broadening of the frequency spectrum of the excitation pulse as its duration is reduced, it is only possible to say that $\Delta t \lesssim \hbar\Delta E^{-1}$.

2.5 Totally Symmetric Modes and Franck-Condon Scattering

The appropriate FC overlap integrals for Raman scattering are of the mixed type: $\langle n + p|v\rangle\langle v|p\rangle$ for the contribution of the p-th hot band to the n-th harmonic ($n = 1$ for the fundamental). These are only non-zero if the orthogonality is removed between a ground-state vibrational level and an excited-state level with a different vibrational quantum number. If, along some particular normal coordinate, the potential-energy curve for nuclear vibration in the excited electronic state can be obtained from that for the ground electronic state merely by a translation along the energy axis by a purely electronic excitation energy, then the orthonormality of the vibrational levels is preserved. In this case the FC integrals give rise only to Rayleigh scattering *even at resonance*. This is the situation referred to in Section 2.4 as the accidental vanishing of the FC terms. The removal of orthogonality and the occurrence of Raman scattering can be attributed in the main to the following two causes: first, a difference of vibrational frequency between the ground and excited states; second, and more effective, a shift in the equilibrium position upon excitation. The first cause is the less effective because the size of the frequency change necessary to give significant values to the appropriate products of FC integrals is larger than is ordinarily encountered. In other words, it is usually a good approximation to assume a common frequency for ground and excited states. Neglecting the effects of anharmonicity, a difference in frequency can only produce non-zero values of $\langle n + p|v\rangle\langle v|p\rangle$ for the n-even harmonics

(*18*), resulting in an alternation of intensities along a progression, the fundamental and its odd harmonics being absent or very weak. This type of behaviour has not been observed, indicating that this mechanism is not commonly of any importance. The difference between the two states with respect to the position of the minimum of the potential curve is zero, by symmetry, along non-totally symmetric coordinates. This shift in equilibrium position may be finite and large [e.g. ~ 0.35 Å for $\tilde{B}\,(^3\Pi^+_{0u}) \leftarrow \tilde{X}(^1\Sigma^+_g)$ of I_2 (*17*)] for totally symmetric modes, giving rise to long overtone progressions. Such progressions are generally not observed for non-totally symmetric modes.

To show the reason for this symmetry restriction, the cause of the shift in equilibrium position will now be examined. Consider a molecule at vibrational equilibrium in its ground electronic state so that there are no net vibrational forces on it. Now let it undergo an electronic transition. If, upon arrival in the excited state, the nuclei are subject to a force, tending to establish some new equilibrium geometry along a particular normal coordinate, then the potential minimum of the excited state is shifted along this coordinate relative to the ground state. (This description obviously does not apply to dissociating excited states with no minima.) The key point, then, is the creation of a force in the excited state, and this must arise from that particular displacement of the electrons which results from the excitation. This nuclear 'following' of an adiabatic electronic motion is a manifestation of one-state vibronic coupling. Since the potential energy of nuclear vibration is equal to the total electronic energy in the ABO approximation, it is possible to write the force along a coordinate, Q, experienced in the electronic excited state, $|e\rangle$, as the diagonal vibronic-coupling matrix element: $\langle e|\partial\mathscr{H}/\partial Q|e\rangle_0$. The force is directly related to the slope of the potential curve in the state $|e\rangle$ evaluated at the ground-state equilibrium configuration, $Q = 0$. If $|e\rangle$ is non-degenerate, the matrix element of the force vanishes unless Q transforms under the totally symmetric irreducible representation of the point group appropriate to $Q = 0$. This is the reason for the rule: shifts of equilibrium position only take place along totally symmetric coordinates.

If $|e\rangle$ transforms under a degenerate irreducible representation, Γ_e (not a damping factor!), then the totally symmetric part of the direct product $\Gamma_e \otimes \Gamma_e$ can still allow a shift of potential minimum, which will be equal for all degenerate components of $|e\rangle$. However, if non-totally symmetric vibrational coordinates are available belonging to irreducible representations contained in the symmetric part, $\{\Gamma_e \otimes \Gamma_e\}$, of the direct product, these coordinates may cause a Jahn-Teller effect in the excited state. For example, if $\Gamma_e = E(D_4)$ then

$$E \otimes E = \{A_1 + B_1 + B_2\} + [A_2]$$

where the curly brackets contain the symmetric part and the square brackets the antisymmetric part of the direct product (*19*). Modes of B_1 and B_2 symmetry will cause equal shifts of opposite sign in the potential minima of the E_x and E_y components. An A_2 mode is not Jahn-Teller active in this case since it preserves the degeneracy of E states, causing only a rotation of the components about the z axis in a descent in symmetry from D_4 to C_4. A_2 modes can, however, cause the scattering tensor to be antisymmetric in the resonance region and thus display inverse polarisation. These latter effects are discussed in Section 2.11.

2.6 The Vibronic View of Franck-Condon Scattering

Returning now to the case of a non-degenerate resonant state, *Tang* and *Albrecht* (*15*) have shown how Albrecht's A term,

$$A = \sum_{e \neq g} \sum_{v} \left[\frac{\varrho^0_{ge} \sigma^0_{eg}}{E_{ev} - E_{gu} - E_L} + \frac{\sigma^0_{ge} \varrho^0_{eg}}{E_{ev} - E_{gw} + E_L} \right] \langle w|v\rangle\langle v|u\rangle \qquad (10)$$

giving the FC contribution to scattering, can give rise to the one-state vibronic-coupling description when the dependence of the energy denominators on the nuclear coordinates is explicitly treated using the vibronic expansion method. This expansion diverges in the resonance region so that the treatment is only valid far from resonance. This is why the A term above has the damping factors omitted for convenience. Despite these restrictions, the method will be briefly outlined here to demonstrate the connection between the overlap and vibronic views of A-term scattering.

The vibronic expansion of the resonance denominator is given by

$$(E_{ev} - E_{gu} - E_L)^{-1} = \sum_{N=0}^{\infty} (-1)^N \frac{(E_{ev} - E^0_e)^N}{(E^0_e - E_{gu} - E_L)^{N+1}}$$

where the superscript zero refers to the equilibrium nuclear configuration of the ground state. Thus, E^0_e is the 'vertical' transition energy which corresponds to the Franck-Condon absorption maximum. Far from resonance, it is sufficient to consider the N = 1 term only, since $(E_{ev} - E^0_e) \ll (E^0_e - E_{gu} - E_L)$. The N = 0 term is independent of v and its contribution to the A term vanishes for a Raman transition (w ≠ u) due to the closure sum rule. Making use of the fact that, in the adiabatic approximation, $E_{ev}|v\rangle = \langle e|\mathscr{H}|e\rangle|v\rangle$, the resonant part of A becomes

$$A' = \sum_{e \neq g} \varrho^0_{ge} \sigma^0_{eg} \sum_{v} \frac{\langle w|\langle e|\mathscr{H} - E^0_e|e\rangle|v\rangle\langle v|u\rangle}{(E^0_e - E_{gu} - E_L)^2}$$

Now the Hamiltonian may be expanded in a Taylor series in Q (the vibrational coordinate of interest) about Q = 0, corresponding to the ground-state equilibrium configuration. Thus,

$$\langle e|\mathscr{H}(Q) - E_e^0|e\rangle = \langle e|\mathscr{H}(0) + \left(\frac{\partial \mathscr{H}}{\partial Q}\right)_0 Q + \ldots - E_e^0|e\rangle$$

$$= \langle e|\mathscr{H}(0)|e\rangle + \langle e\left|\frac{\partial \mathscr{H}}{\partial Q}\right|e\rangle_0 Q + \ldots - E_e^0$$

$$= \langle e\left|\frac{\partial \mathscr{H}}{\partial Q}\right|e\rangle_0 Q$$

as far as the linear term, since $\langle e|\mathscr{H}(0)|e\rangle = E_e^0$. The final result is therefore

$$A' = \sum_{e \neq g} \varrho_{ge}^0 \sigma_{eg}^0 \sum_v \frac{\langle w|Q|v\rangle \langle e\left|\frac{\partial \mathscr{H}}{\partial Q}\right|e\rangle_0 \langle v|u\rangle}{(E_e^0 - E_{gu} - E_L)^2} \tag{11}$$

where for a $0 \rightarrow 1$ fundamental the vibrational part reduces to $\langle 1|Q|0\rangle$ by closure over v. If the vibrational potential energy of the excited state is written as

$$V_e(Q) = V_e(0) + \left(\frac{\partial V_e}{\partial Q}\right)_0 Q + \frac{1}{2}\left(\frac{\partial^2 V_e}{\partial Q^2}\right)_0 Q^2 \tag{12}$$

then the linear term is associated with the force, $F_e^0 = (\partial V_e/\partial Q)_0$, experienced in the excited state. The term $(\partial^2 V_e/\partial Q^2)_0 = (\partial F_e^0/\partial Q)_0 = k_e$, represents the force constant in the excited state. Higher derivatives need only be considered for cubic or higher potentials; in the harmonic approximation they are obviously zero. Putting these interpretations upon Eq. (12) gives

$$V_e(Q) = V_e(0) + \langle e\left|\frac{\partial \mathscr{H}}{\partial Q}\right|e\rangle_0 Q + \frac{1}{2} k_e Q^2$$

$$= V_e(0) + \frac{1}{2} k_e \left[Q + \frac{\langle e\left|\frac{\partial \mathscr{H}}{\partial Q}\right|e\rangle_0}{k_e}\right]^2 - \frac{1}{2}\frac{\langle e\left|\frac{\partial \mathscr{H}}{\partial Q}\right|e\rangle_0^2}{k_e} \tag{13}$$

If this expression is compared with that appropriate to a simple, displaced, one-dimensional harmonic oscillator:

$$V_e(Q) = V_e(0) + \frac{1}{2} k_e(Q + \delta)^2$$

then the vibronic-coupling matrix element may be related to the displacement, δ, between the ground and excited state potential minima by the simple formula

$$\langle e \left| \frac{\partial \mathscr{H}}{\partial Q} \right| e \rangle_0 = k_e \, \delta \, . \tag{14}$$

Thus, the oscillator described in Eq. (13) is not only shifted along Q by the amount δ, but it also has its potential curve depressed by the quantity $\frac{1}{2} k_e \delta^2$.

Equation (14) allows a translation between the vibrational overlap picture and the vibronic coupling view. The overlap integrals are calculated in terms of the displacement, δ; for example, using the formula given by *Krushinskii* and *Shorygin* 1961 (*18*):

$$\langle u \mid v \rangle = \left(\frac{u!}{v!}\right)^{1/2} \exp(-\alpha\delta^2/4) \left(\sqrt{\frac{\alpha}{2}}\,\delta\right)^{v-u} \mathscr{L}_v^{v-u}(\alpha\delta^2/2)$$

for $u \leqslant v$. $\alpha = k_e/\hbar\omega$ and $\mathscr{L}$ is a Laguerre polynomial.

In the vibronic view the relevant parameter is the vibronic coupling strength $\Delta_{01}^{e} = \langle e \mid \partial\mathscr{H}/\partial Q \mid e \rangle_0 \langle 0 \mid Q \mid 1 \rangle$. In general,

$$\Delta_{v-1,v}^{e} = \Delta_{v,v-1}^{e} = \sqrt{v}\, \Delta_{01}^{e}$$

since

$$\langle v \mid Q \mid v-1 \rangle = \sqrt{v} \left(\frac{\hbar\omega}{2k_e}\right)^{1/2} = \sqrt{v}\, \langle 1 \mid Q \mid 0 \rangle \tag{15}$$

where the harmonic force constant is given by $k_e = \omega^2 \mu = 4\pi^2 c^2 \nu^2 \mu$ and μ is the appropriate reduced mass.

The relation between the two approaches has been described by *Hong* 1977 and 1978 (*20*). However, in calculating the Raman transition probabilities in the vibronic view, *Hong* made use of the relatively unfamiliar Green's-function method. A more conventional derivation of the essential results will be given here.

Hong's methods, which are a general extension of the work of *Nafie et al.* (*21*), make use of the Longuet-Higgins representation for vibronic coupling calculations (*22*). In this basis it is assumed that a complete set of 'true' vibrational states is available for the ground electronic state. For convenience, these are assumed to be harmonic. The electronic parts of the BO products are all evaluated at the ground-state equilibrium configuration (the crude BO approximation). The product of an excited electronic CBO state with the set of vibrational states mentioned forms the basis within which the effects of excited-state vibronic coupling are calculated. This initial set of basis states is represented by the levels of the dashed potential curve in Fig. 4B. The effect of the vibronic perturbation, $\Delta_{v,v\pm1}^{e}$, which is of the first order in the vibrational coordinate, is to mix only the neighbouring levels v_e, $(v \pm 1)_e$. Thus $|0_e)$ interacts with $|1_e)$, $|1_e)$ interacts with $|0_e)$ and $|2_e)$, and so on. Since the zero-order vibrational states (round-bracketed kets) are independent of the electronic state in the Longuet-Higgins basis, the subscript, e, may be dropped. The diagonalisation of the complete vibronic coupling matrix in the CBO electronic state, $|e\rangle$, would lead to the set of levels shown in the full upper-potential curve of Fig. 4B. These vibronic levels possess an origin shifted relative to that of the ground state by the amount $\delta = \langle e|\partial\mathscr{H}/\partial Q|e\rangle_0/k_e$ and an energy depression (relative to the zero-order) of $\frac{1}{2}k_e\delta^2 = \langle e|\partial\mathscr{H}/\partial Q|e\rangle_0^2/2k_e$.

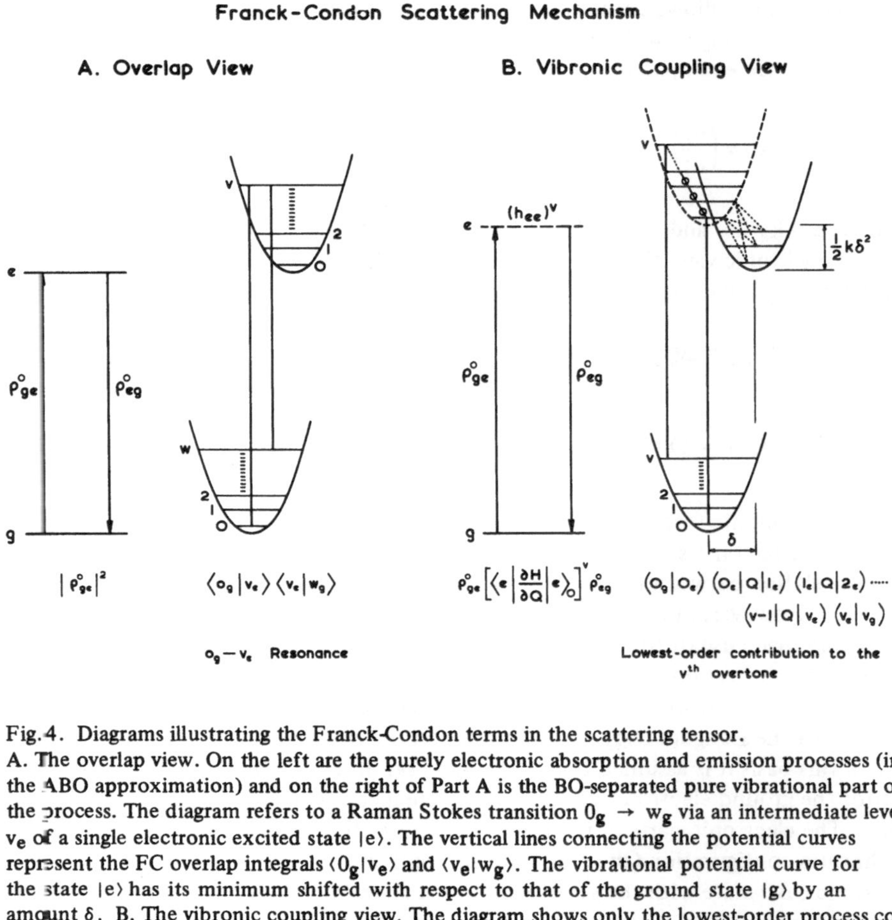

Fig. 4. Diagrams illustrating the Franck-Condon terms in the scattering tensor.
A. The overlap view. On the left are the purely electronic absorption and emission processes (in the ABO approximation) and on the right of Part A is the BO-separated pure vibrational part of the process. The diagram refers to a Raman Stokes transition $0_g \rightarrow w_g$ via an intermediate level v_e of a single electronic excited state $|e\rangle$. The vertical lines connecting the potential curves represent the FC overlap integrals $\langle 0_g|v_e\rangle$ and $\langle v_e|w_g\rangle$. The vibrational potential curve for the state $|e\rangle$ has its minimum shifted with respect to that of the ground state $|g\rangle$ by an amount δ. B. The vibronic coupling view. The diagram shows only the lowest-order process contributing to the production of the vth harmonic (v = 1 for a fundamental). The basis states for the evaluation of vibronic coupling in the excited state $|e\rangle$ are represented by the levels of the dashed potential curve, which is identical to the ground-state potential apart from a purely electronic excitation energy. The diagonalisation of the complete vibronic-coupling matrix would lead to a set of shifted and depressed levels represented by the full upper curve. The dashed lines connecting the levels of the two upper curves represent some of the contributions from the initial basis functions to the composition of the final vibrational eigenstates. Overlaps appear in the diagram as vertical lines, and matrix elements of the vibrational coordinate operator, Q, as circled lines connecting the excited levels. h_{ee} is an abbreviation for the electronic vibronic-coupling matrix element $\langle e|\partial\mathscr{H}/\partial Q|e\rangle_0$

This complete diagonalisation is unnecessary for a consideration of Raman scattering since the system returns finally to the ground state again. Therefore, in the vibronic view, only those parts of the interaction which are necessary to produce a given overtone are considered; this is done conveniently using the Green's function formalism (*20*). The Raman process illustrated in Fig. 4B may be described by the following formal sequence of events: A BO-separated pure electronic excitation takes the system into the excited state via the electronic transition moment $\boldsymbol{\varrho}_{ge}^{0}$ and a diagonal vibrational overlap $(0_g|0_e)$. The first-order vibronic coupling interaction then occurs v times, creating v vibrational quanta, before the system returns to the ground electronic state via the transition moment $\boldsymbol{\varrho}_{eg}^{0}$ and the diagonal overlap $(v_e|v_g)$.

If the interaction of only the $|0)$ and $|1)$ levels in the CBO state $|e\rangle$ is considered in some detail, the general results should be made clear. The resultant first-order perturbed levels are given by

$$\left.\begin{aligned} |0_e\rangle &= |0) - \Delta_{01}^{e}(E_1 - E_0)^{-1}|1) \\ |1_e\rangle &= \Delta_{01}^{e}(E_1 - E_0)^{-1}|0) + |1) \end{aligned}\right\} \qquad (16)$$

where $E_v = E_{e0} + v\hbar\omega + i\,\Gamma_{ev}$ and so $E_1 - E_0 = \hbar\omega$.

The correction to the energy of the zeroth level is given by

$$\langle 0_e|\mathscr{H}|0_e\rangle - (0_e|\mathscr{H}|0_e) = -\frac{(0_e|\mathscr{H}|1_e)^2}{E_1 - E_0} = -\frac{(\Delta_{01}^{e})^2}{\hbar\omega} = -\frac{1}{2}k_e\,\delta^2$$

since

$$\Delta_{01}^{e} = \langle e\left|\frac{\partial\mathscr{H}}{\partial Q}\right|e\rangle_0\,(0|Q|1) = k_e\,\delta(\hbar\omega/2\,k_e)^{1/2}\,.$$

The Franck-Condon A term will now be written in the overlap picture for the case of fundamental scattering in the low temperature limit where excitation occurs only from the $|0)$ level of the ground state. It must be remembered that the ground levels are 'true' in the zero-order, i.e. there is no vibronic coupling in the ground state.

$$(\boldsymbol{\alpha}_{\rho\rho})_{0\to 1} = |\boldsymbol{\varrho}_{ge}^{0}|^2\left[\frac{(0|0_e\rangle\langle 0_e|1)}{E_0 - E_L} + \frac{(0|1_e\rangle\langle 1_e|1)}{E_1 - E_L}\right] \qquad (17)$$

Substituting Eqs. (16) for the first-order perturbed intermediate vibronic levels and making use of the orthonormality of the zero-order states $[(0|0) = (1|1) = 1; (0|1) = (1|0) = 0]$, the Franck-Condon factors become

$$(0|0_e\rangle\langle 0_e|1) = -\frac{(0|0)\,\Delta_{01}^{e}(1|1)}{E_1 - E_0} = -\frac{\Delta_{01}^{e}}{E_1 - E_0}$$

$$(0|1_e\rangle\langle 1_e|1) = \frac{(0|0)\,\Delta_{01}^{e}(1|1)}{E_1 - E_0} = \frac{\Delta_{01}^{e}}{E_1 - E_0}$$

When these expressions are substituted into Eq. (17) the result is

$$(\alpha_{\rho\rho})_{0\to 1} = |\varrho^0_{ge}|^2 \frac{\Delta^e_{01}}{E_1 - E_0}\left[\frac{1}{E_1 - E_L} - \frac{1}{E_0 - E_L}\right]$$

$$= |\varrho^0_{ge}|^2 \frac{\Delta^e_{01}}{(E_0 - E_L)(E_1 - E_L)} .$$

In Green's function notation this would be written

$$(\alpha_{\rho\rho})_{0\to 1} = \varrho^0_{ge}\, g^e_{00}\, \Delta^e_{01}\, g^e_{11}\, \varrho^0_{eg}$$

where the zero-order Green's functions are defined by *(20)* $g^e_{nn} = (E_n - E)^{-1}$ and E is for our purposes the energy, E_L, of the exciting radiation. This result represents the lowest-order contribution to the fundamental. In general the lowest-order contribution to the n-th harmonic is of the n-th order and for each value of n there is only one term of this order, corresponding to a unique path between initial and final states *(20)*. (n = 1 for the fundamental.) For example, the third-order contribution to the second overtone is given by

$$(\alpha_{\rho\rho})_{0\to 3} = \varrho^0_{ge}\, g^e_{00}\, \Delta^e_{01}\, g^e_{11}\, \Delta^e_{12}\, g^e_{22}\, \Delta^e_{23}\, g^e_{33}\, \varrho^0_{eg}$$

$$= |\varrho^0_{ge}|^2 \frac{\sqrt{2}\sqrt{3}\,(\Delta^e_{01})^3}{(E_0 - E_L)(E_1 - E_L)(E_2 - E_L)(E_3 - E_L)}$$

using Eq. (15).

This type of process for the production of a general overtone is illustrated in Fig. 4B. In general

$$(\alpha_{\rho\rho})_{0\to n} = |\varrho^0_{ge}|^2 \frac{\sqrt{n!}\,(\Delta^e_{01})^n}{\prod_{m=0}^{n}(E_m - E_L)} \qquad (18)$$

The ratio of the intensity of the n-th harmonic to that of the fundamental is given by

$$R_n = \frac{I(n\nu)}{I(\nu)} = \frac{n}{2}\left[\frac{(\nu_L - n\nu)^4\,(\nu_L - \nu)^4}{(\nu_L - 2\nu)^4\,\{\nu_L - (n-1)\,\nu\}^4}\right]\left[\frac{(2\nu)^2 + \Gamma_e^2}{(n\nu)^2 + \Gamma_e^2}\right] R_2\, R_{n-1} \qquad (19)$$

for resonance with the zeroth vibrational level of the excited state, i.e. $\nu_L = (E_{e0} - E_{g0})/hc$. The half band-width, Γ_e, is taken to be independent of the vibrational quantum numbers in the electronic state $|e\rangle$.

These lowest-order contributions were used by *Nafie et al.* *(16)* to calculate the relative intensities of the series of overtones observed in the resonance Raman spectrum of iodine vapour (see Section 4.1). The calculated intensities were found to fall off more rapidly with increasing overtone number than was observed experimentally, being low by a factor of ~ 2000 by the 15th overtone. This behaviour indicates the inadequacy of the lowest-order treatment, which is only strictly applicable in cases of weak vibronic coupling with $(\Delta^{e}_{01}/\hbar\omega) \lesssim 1$, and the contributions to the n-th harmonic from perturbation terms of order n + 2, n + 4, etc. must be included. Whereas there is only one term of the n-th order for an n-th harmonic, there are (for example) n + 1 terms of order n + 2 and this correction will therefore favour the higher overtones. These higher-order contributions are included in *Hong's* treatment (*20*) in the form of 'self energy' corrections to the Green's functions. If the self energies are put equal to zero in *Hong's* equations the weak-coupling results of *Nafie* are obtained.

Both the weak and the strong coupling treatments make use of a harmonic basis and it is therefore inappropriate to apply them to the case of continuum scattering, as attempted by *Nafie et al.* 1971. Even in discrete scattering (below the convergence limit), strong coupling (large potential shift) in real molecules is inseparable from the effects of anharmonicity since the FC maximum is shifted towards the convergence limit.

2.7 *Excitation Profiles for Totally Symmetric Modes*

A characteristic feature of scattering for totally symmetric modes is that contributions arise generally from all the vibrational levels of the resonant electronic state. The A-term scattering intensity depends on the square modulus of the scattering tensor, which is given by (resonant part only)

$$|\boldsymbol{\alpha}_{0\to n}|^2 = \Big[\sum_v \boldsymbol{\alpha}(v)\Big]\Big[\sum_v \boldsymbol{\alpha}(v)\Big]^*$$

$$= |\rho^0_{ge}|^2 \left[\sum_v \frac{|\langle 0|v\rangle\langle v|n\rangle|^2}{E_v^2 + \Gamma_v^2} + 2\sum_{v<v'}\sum \frac{\langle 0|v\rangle\langle v|n\rangle\langle 0|v'\rangle\langle v'|n\rangle(E_v E_{v'} + \Gamma_v \Gamma_{v'})}{(E_v^2 + \Gamma_v^2)(E_{v'}^2 + \Gamma_{v'}^2)}\right] \tag{20}$$

where $E_v = E_{e0} - E_{g0} + v\hbar\omega - E_L$.

The cross terms appearing in the double summation of Eq. (20) represent the interference between the contributions from different levels v, v′. This may lead to an increase or decrease in scattering intensity compared with that resulting from the sum of individual resonances [first sum in Eq. (20)]. Destructive interference is possible because, when the Raman FC factors are individually non-zero, the closure sum rule $(\sum_v \langle 0|v\rangle\langle v|n\rangle = 0$ for $n \neq 0)$ implies that, for some values of v, $\langle 0|v\rangle\langle v|n\rangle$ must be negative. The relative importance of the various terms of Eq. (20) depends markedly on the appearance of the absorption band. If the bandwidths are much smaller than the vibrational intervals $(\Gamma_v \ll \hbar\omega = E_{v+1} - E_v)$ then discrete structure is seen in absorption, as for example in gas phase spectra of diatomic molecules. Only a brief account will be given here of the essential features of the different types of scattering and the reader is referred to a recent review (*23*) and references therein for more details. In the case of a diatomic molecule, if excitation occurs within a discrete absorption v ← 0, then $E_v^2 = 0$ and one squared term predominates in Eq. (20). The other squared terms and the cross terms are negligible since $\Gamma_v^{-2} \gg (\hbar^2 \omega^2 + \Gamma_{v'}^{-2})^{-1}$. This is the so-called resonance fluorescence limit which is also referred to as discrete resonance-Raman scattering. The scattering intensity as a function of v or n will, in general, behave irregularly, depending on the particular values of the Raman FC integrals $\langle 0|v\rangle\langle v|n\rangle$. In this type of scattering the lifetimes of intermediate states are long and the scattering is susceptible to quenching by increased gas pressure. The latter effect is due to a collisional contribution to the Γ_v. When rotational substructure is visible in absorption, discrete resonance Raman scattering occurs with excitation within a rotational component and the description of the intermediate state must be supplemented with the rotational quantum number J. The separation of the rotational levels is usually sufficiently small to give significant contributions from transitions which are only slightly off resonance.

An interesting consequence of the long lifetimes for discrete resonance is that the molecule may perform numerous rotations before re-emission. This causes depolarisation of the resonance-scattered radiation. For totally-symmetric vibrational modes in the normal Raman effect the scattering is polarised since the lifetime of the intermediate state is very short compared with a rotational period ($\sim 10^{-11}$ s).

Remaining within a region of discrete structure, if excitation now occurs between absorption lines it is still only slightly off resonance with some rotational transitions. Thus the lifetimes of these virtual states are long enough to allow depolarisation of the scattered radiation. The lifetimes are, however, less than, or of the same order of magnitude as, the mean free time so that very little quenching is observed. In the latter respect this slightly off-resonance scattering is like normal Raman scattering and unlike resonance fluorescence. This situation is sometimes referred to as preresonance Raman scattering. The large depolarisation observed for this type of scattering in diatomics [e.g. for I_2 in Ref. (*24*)] is consistent with only a small number of rotational states being important since a summation over many states leads to polarised scattering as a result of closure over the rotational states.

When excitation occurs above the dissociation limit of the resonant state (continuum resonance Raman), the scattering times are small, as in normal Raman scattering, being limited by the time needed for the atoms to fly apart ($\sim 10^{-12}$ s). The scattering intensity is therefore not quenched by addition of foreign gases and depolarisation is small. The continuum nature of the excited state means that, for a particular excitation energy, every populated vibrational-rotational level of the ground-state is in resonance with some level of the continuum. The observed scattering is therefore a superimposition of these various contributions. In addition, there are many nearby states in the continuum which are only slightly off resonance and which therefore make significant contributions to the scattering. These properties of the continuum states have an averaging effect such that the scattered intensity is a smoothly varying function of overtone number or excitation energy, in contrast to the discrete case.

The diverse behaviour which has been described for discrete and continuum scattering is a quality of small molecules in the gas phase, exemplified by the halogens. Many polyatomic inorganic ions, in solution or in the solid phase, display a variable degree of vibronic structure in their electronic absorption bands. This depends on the nature of the ion and of the electronic transition, as well as on the temperature and state of aggregation. With greater than two atoms there is the possibility of more than one totally symmetric mode giving rise to secondary FC progressions, as well as progressions based on combination bands (false origins). These effects, together with the many mechanisms which may cause broadening of the discrete structure in condensed phases, can result in a quasi-continuum of overlapping and broadened levels. Excitation profiles are expected to show vibronic structure to the extent that this is discernable in the absorption profile. However, the excitation profiles have the advantage that the scattered intensity may be monitored individually for different fundamentals and combination bands and this can, in principle, allow the resolution of component vibronic structure in a complex absorption band. The situation is complicated in the following way. When the bandwidths are not much smaller than the vibrational intervals ($\Gamma \lesssim \hbar\omega$) the cross terms of Eq. (20) are of comparable magnitude to the squared terms and considerable interference results. Destructive interference causes the excitation profiles to peak differently from the absorption profile and this is particularly so for the higher overtones. Excitation profiles of this type have been calculated for the visible absorption bands of the manganate (*26*) and permanganate (*27*) ions although the experimental data are far from complete.

With large bandwidths, $\Gamma \gg \hbar\omega$, the absorption profile is smooth, showing no vibronic structure, and the excitation profiles appear similar. The excitation profile of a fundamental will usually peak near the FC maximum of the absorption band; if not, this may indicate more than one electronic transition under the same absorption band contour. An increase in the bandwidths with increasing vibrational quantum number in the excited state can lead to an excitation-profile peak red-shifted relative to the absorption band maximum (*25*). Even with smooth absorption bands the overtone profiles need not peak at the same energy as that of the fundamental and Eq. (20) predicts a progressive blue shift (each shift $\lesssim \hbar\omega$) with increasing overtone number. This has been observed for the ν_1 bands of Wolffram's red and the $[FeBr_4]^-$ ion (*28, 29*) (Section 4).

Considerable modifications to the appearance of excitation profiles may result from interference between the scattering contributions from more than one electronic transition. Destructive interference may lead to a decrease in scattering intensity in the region of an absorption band. This type of interference has been observed between the weak resonance of a Laporte-forbidden d–d transition and the preresonance scattering from a higher-energy allowed transition in the $[Co(en)_3]^{3+}$ ion (*31,32*). The possibility of similar effects in the region of the forbidden $^1B_{2u}$ band of benzene has been discussed recently (*33*).

2.8 *Combination-Band Progressions Involving Solely Totally Symmetric Modes*

When a long overtone progression is observed for a totally symmetric mode in a resonance Raman spectrum, it is not uncommon to see secondary progressions in the frequency of this mode, based upon one quantum of another totally symmetric mode. The members of the secondary progression may be designated by $v_1 \nu_1 + \nu_2$ where $v_1 = 1, 2$, etc.; ν_1 is the progression-forming mode and ν_2 is the enabling mode. This behaviour is found, for example, in the $[Mo_2X_8]^{4-}$ ions where ν_1 is the Mo–Mo stretching frequency and ν_2 is the Mo–X stretching frequency (see Section 4.5), both modes possessing $A_{1g}(D_{4h})$ symmetry. Another example is found in the mixed-valence linear chain compounds, e.g. Wolfram's Red $[Pt^{IV}(etn)_4Cl_2][Pt^{II}(etn)_4]Cl_4 \cdot 4H_2O$ (Section 4.7), where progressions in the symmetric Pt–Cl stretching frequency are based upon totally symmetric fundamentals of the ethylamine ligands. Progressions in totally symmetric modes based on non-totally symmetric fundamentals are less common and these are discussed in Section 2.10.

These secondary progressions may be discussed in terms of the general two-dimensional Franck-Condon factors $\langle 0_1\, 0_2 | n_1\, n_2 \rangle \langle n_1\, n_2 | v_1\, v_2 \rangle$. The numerical subscripts to the vibrational quantum numbers refer to the two coordinates Q_1, Q_2 with fundamental frequencies ν_1, ν_2 respectively. In the independent-mode approximation, these FC factors are directly factorisable, i.e. $\langle 0_1\, 0_2 | n_1\, n_2 \rangle \langle n_1\, n_2 | v_1\, v_2 \rangle = \langle 0_1\, n_1 \rangle \langle n_1 | v_1 \rangle \langle 0_2 | n_2 \rangle \langle n_2 | v_2 \rangle$, and thus may be evaluated in terms of the one-dimensional integrals for the individual modes. The independent-mode approximation breaks down if there are significant changes in the composition of the normal coordinates as a result of force-constant changes (diagonal and non-diagonal) resulting from the electronic excitation. This is the Duschinsky effect (*34*) in which the new excited-state normal coordinates are generally some linear combination of the ground-state coordinates. Thus a vibrational level corresponding to one of the new excited-state coordinates may have overlaps with the levels of both ground-state coordinates. A significant Duschinsky effect might be expected when the change in equilibrium geometry upon excitation has components of large and comparable magnitude along two totally symmetric (ground-state) coordinates. In this case the resonance Raman spectrum should show long progressions in both modes together with high-order combinations $v_1 \nu_1 + v_2 \nu_2 (v_1, v_2 > 1)$. In the quoted examples, however, $v_2 = 1$ only, suggesting a small component of geometric change along Q_2, which is confirmed by the absence of overtones of ν_2. This implies that the only important intermediate levels will have $n_2 = 0$ or 1.

The two-dimensional FC approach brings out the point that, in general, a particular Raman band, attributable to a totally symmetric mode, obtains intensity enhancement by excitation in resonance with transitions to each of the combination levels $| n_1\, n_2 \rangle$ of the excited state. This is true even for the individual fundamentals through terms of the type

$$\langle 0_1\, 0_2 | n_1\, n_2 \rangle (E_0 + n_1 \hbar\omega_1 + n_2 \hbar\omega_2 - E_L)^{-1} \langle n_1\, n_2 | 1_1\, 0_2 \rangle \qquad (21)$$

where $E_0 = E_{e0} - E_{g0} + i\,\Gamma_{en_1,n_2}$. In addition, a fundamental of one mode may obtain a contribution to its intensity from resonance with a hot-band transition of another mode:

$$\langle 0_1\, 1_2 | n_1\, n_2 \rangle (E_0 + n_1\, \hbar\omega_1 + (n_2 - 1)\, \hbar\omega_2 - E_L)^{-1} \langle n_1\, n_2 | 1_1\, v_2 \rangle \quad (22)$$

where $v_2 = 1$ for the pure ν_1 fundamental transition. If $v_2 = 0$ here, this gives rise to a difference band $\nu_1 - \nu_2$. If the ground-state levels 1_1 and 1_2 have comparable populations at the temperature of the experiment, it is conceivable that a difference band, $\nu_1 - \nu_2$ could arise where $\nu_1 < \nu_2$ and therefore the band would be seen in the anti-Stokes region of the spectrum. Possible candidates for this type of difference band have been pointed out in the resonance Raman spectrum of cytochrome c by *Friedman* and *Hochstrasser* 1973 (*35*). Although in their case the Raman bands arise from non-totally symmetric modes and a second-order Herzberg-Teller mechanism (Section 2.9) is employed, this merely restricts the resonant combination levels to those of the type $|0_1\, 1_2\rangle$; the principle is the same.

The appearance, in progressions of the type $v_1\, \nu_1 + \nu_2$, of only a single quantum of the mode, Q_2, suggests an alternative explanation. The geometric change on resonant excitation may be zero along Q_2 but this mode may be active in HT vibronic coupling with a nearby transition of the same symmetry as the resonant electronic transition. Mixed A and B term scattering results:

$$\begin{aligned}(\alpha_{\rho\sigma})_{0_1\, 0_2 \to v_1\, 1_2} &= \sum_{n_1} \langle v_1 | n_1 \rangle \langle n_1 | 0_1 \rangle [\boldsymbol{\varrho}^0_{gs}\, h_{se}\, \boldsymbol{\sigma}^0_{eg} \langle 1_2 | Q_2 | 0_2 \rangle \langle 0_2 | 0_2 \rangle (E_0 + n_1\, \hbar\omega_1 - E_L)^{-1} \\ &\quad + \boldsymbol{\varrho}^0_{ge}\, h_{es}\, \boldsymbol{\sigma}^0_{sg} \langle 1_2 | 1_2 \rangle \langle 1_2 | Q_2 | 0_2 \rangle (E_0 + n_1\, \hbar\omega_1 + \hbar\omega_2 - E_L)^{-1}]\end{aligned} \quad (23)$$

where $h_{es} = \langle e \left| \frac{\partial \mathscr{H}}{\partial Q_2} \right| s \rangle (E_e - E_s)^{-1}$. The first term in square brackets gives resonances with the main ν_1-band progression in the absorption spectrum and the second term gives resonances with the combination-band progression $n_1\, \nu_1 + \nu_2$. Since, in the pure FC mechanism, a small geometric change along Q_2 restricts n_2 to the values 0 and 1 only (consistent with the absence of overtones of ν_2) and produces the same resonance conditions as above, it is difficult to distinguish between the two mechanisms. In high symmetries where the transition moments of different polarisation transform distinctly, Eq. (23) must have $\rho = \sigma$ if Q_2 is totally symmetric. However, in low symmetries where non-parallel transition moments may transform under the same irreducible representation (point groups C_1, C_2, C_i, C_s and C_{2h}), non-diagonal tensor components may result from HT coupling via totally symmetric modes. In these cases the coupling may produce measurable effects on the depolarisation ratio for an active mode, as pointed out by *Zgierski* 1976 (*36*).

In the case of the linear-chain compounds mentioned earlier, the secondary progressions have been discussed qualitatively assuming that the FC mechanism operates (*37*). The appearance of ligand modes was attributed to relaxations in the steric interactions between the halide and amine ligands in the intervalence charge-transfer excited state. In the example of the $[Mo_2X_8]^{4-}$ ions, the appearance of the combination-band progression in ν_1(Mo–Mo) based on ν_2(Mo–X) (for excitation resonant with the "$\delta^* \leftarrow \delta$" transition) may be explained in terms of the involvement of the d_{xy} orbitals of the Mo centres in Mo–X π-bonding as well as the axial Mo–Mo δ-bonding. Thus, instead of the simple $\delta^*(b_{1u}) \leftarrow \delta(b_{2g})$ designation (see Section 4.5), two possible transitions of symmetry $b_{1u} \leftarrow b_{2g}$ arise. These may be approximately described as

$$\left.\begin{array}{l}\delta^*(\text{Mo–Mo})\ \pi(\text{Mo–X})\ \pi^*(\text{X–X}') \leftarrow \\ \delta^*(\text{Mo–Mo})\ \pi^*(\text{Mo–X})\ \pi^*(\text{X–X}') \leftarrow\end{array}\right\}\delta(\text{Mo–Mo})\ \pi(\text{Mo–X})\ \pi(\text{X–X}')$$

X and X′ refer to halogen ligands on different Mo centres. The second transition is expected at higher energy than the first, since π(Mo–X) interaction is very likely greater than the δ(Mo–Mo) interaction). It seems likely that the absorption bands observed at around 18,000 cm^{-1} in $[Mo_2X_8]^{4-}$ ions X = Cl, Br (Section 4.5) correspond essentially to the first transition shown above and that the involvement of the ν_2(Mo–X) fundamental in the observed resonance Raman spectra results from small admixtures of the second transition. This mixing may occur in the zero-order, via non-vibronic perturbations, since the relevant states are of the same symmetry. In this case the FC description is appropriate. With excitation in resonance with the second (presumably higher-energy) transition it is expected that the resonance Raman spectrum will be dominated by a progression in the ν_2(Mo–X) mode, since the transition is predominantly $\pi^* \leftarrow \pi$(Mo–X) in character. This type of analysis is an example of the general rule stated by *Tsuboi* (*38*) and based on the behaviour of the FC factors for totally symmetric modes: viz. the modes enhanced in the resonance Raman spectrum are those responsible for transforming the equilibrium geometry from that of the ground state to that appropriate to the resonant excited state. The cases considered in this section are of interest in exemplifying geometric changes with components along more than one ground-state normal coordinate. This behaviour is also found in some bent triatomic molecules where changes in the equilibrium values of both bond length and bond angle may occur. The qualitative relation between geometry and electronic structure and spectroscopy of triatomic molecules was considered by *Walsh* 1953 (*39*). Combination bands and progressions involving the stretching and bending coordinates are observed in the resonance Raman spectra of NO_2 and ClO_2 (see Section 4.2). The case of NO_2 is complicated by the contribution of several overlapping electronic transitions to the visible absorption spectrum and the resonance Raman results are consistent with geometric changes of comparable magnitude for both the bond angle and the bond lengths. For excitation in the region of the $\tilde{A}\ ^2A_2 \leftarrow \tilde{X}\ ^2B_1$ transition of ClO_2, the main resonance-Raman progression involves the ν_1(Cl–O) symmetric stretch with a weaker secondary progression extending to $2\nu_1 + \nu_2$ where ν_2 is the bending-mode fundamental.

2.9 Non-Totally Symmetric Modes and Herzberg-Teller Scattering

Herzberg-Teller scattering refers to the scattering which arises from the first-order terms in the vibronic expansion of the polarisability [the second and third terms of Eq. (6) in Section 2.2] when these terms are interpreted as arising explicitly from a first-order HT coupling between two different electronic excited states. If these two states are of the same symmetry, they may be coupled through totally symmetric modes and this may be an important source of scattering intensity if the A-term contribution to these modes is small. A-term scattering may be small if there is little or no change in equilibrium geometry upon excitation, leading to small values of the Franck-Condon (FC) overlap integrals. The latter occurrence is quite common in polyatomic molecules and is the reason why they may not show long overtone progressions in totally symmetric modes. The significance of the HT contribution to scattering by totally symmetric modes is made less by the rare proximity of electronic states of the same symmetry, due to the non-crossing rule. However, this mechanism may be relevant in cubic symmetries where, of necessity, all allowed electronic transitions have the same symmetry.

The major importance of the Herzberg-Teller mechanism is the intensity it allows to scattering by non-totally symmetric modes. These modes generally have no *A*-term contribution due to the zero displacement of excited-state potential minima along non-totally symmetric coordinates in non-degenerate states (Section 2.5). However, in a degenerate excited state, a strong Jahn-Teller distortion along an appropriate non-totally symmetric coordinate may lead to significant FC overlaps and an A-term contribution which is comparable to that of the B-term. The B-term scattering arises from first, the Jahn-Teller coupling (a special case of HT coupling) between the different components of the degenerate state [called intramanifold coupling in Ref. (*40*)], and second, the HT coupling between the Jahn-Teller distorted state and other distinct electronic manifolds [intermanifold coupling (*40*)].

The classic cases of the HT mechanism concern coupling between two electronic states of different symmetry or between the different components of two degenerate states of the same symmetry. An important example of the first case occurs when electric dipole transitions to one of the two states are forbidden (e.g. the Laporte-forbidden d–d and f–f transitions). In this case, the forbidden transition may acquire absorption intensity by HT mixing with an allowed transition via a non-totally symmetric mode of appropriate symmetry (the irreducible representation of the active mode must be contained in the direct product of the irreducible representations for the two states coupled by the HT mechanism).

In the second case above, since the states have the same symmetry, both are required to give electric-dipole-allowed transitions. The most important examples of this type, in the context of resonance Raman scattering, are the much studied metallo-porphyrin molecules which constitute the active sites of the haem proteins, notably haemoglobin and cytochrome c (Section 4.8). The visible and near ultraviolet absorption spectra of these systems show two $\pi^* \leftarrow \pi$ transitions of the porphyrin ring; both are allowed with in-plane polarisation and E_u excited-state symmetry. The lower energy Q band is the weaker and consists of an electronic origin (0–0 transition), the Q_0 or α band, in the 500–600 nm region together with a vibronic sideband (Q_1 or β band) maximising some 1300 cm^{-1} higher. The Q_1 sideband arises from HT mixing with the stronger Soret (γ or B) band near 400 nm. The Q_1 band is composite, consisting of the 0–1 transitions arising from all the HT-active modes. The appearance of only the 0–1 sideband indicates the adequacy of a first-order HT approach for the absorption process, i.e. this is a case of weak vibronic coupling. The absence of extended progressions in totally symmetric modes suggests that there is negligible change in equilibrium geometry resulting from the Q transition.

The HT coupling mechanism was first considered in the context of the Raman scattering tensor by *Albrecht* (*9*) whose approach was briefly discussed in Section 2.4.

The first-order terms from Eq. (6) are given in the low temperature limit ($u_g = 0$) for the case of a single non-totally symmetric harmonic mode:

$$\begin{aligned}(\boldsymbol{\alpha}_{\rho\sigma})_{0\to1} = {} & \boldsymbol{\varrho}'_{ge}\,\boldsymbol{\sigma}^0_{eg}\,\langle 1|Q|0\rangle\langle 0|0\rangle(E_{e0} - E_{g0} - E_L + i\,\Gamma_{e0})^{-1} \\ & + \boldsymbol{\varrho}^0_{ge}\,\boldsymbol{\sigma}'_{eg}\,\langle 1|1\rangle\langle 1|Q|0\rangle(E_{e1} - E_{g0} - E_L + i\,\Gamma_{e1})^{-1}\end{aligned} \tag{24}$$

This represents the resonant part only for a single intermediate electronic state $|e\rangle$. For a non-totally symmetric mode, there is no displacement of the potential curve minimum in the excited state (neglecting Jahn-Teller effects). Thus, assuming negligible difference in vibrational frequency between ground and excited states, the vibrational levels remain orthonormal and so only diagonal overlap integrals appear. The restriction of the sum over v_e in Eq. (6) to the values 0 and 1 only in Eq. (24) results from the above assumptions together with the harmonic oscillator selection rule on the matrix elements of Q. Equation (24) may be rewritten employing explicitly the HT coupling expressions for $\boldsymbol{\varrho}'$, $\boldsymbol{\sigma}'$ given by Eq. (9) of Section 2.4:

$$\boldsymbol{\varrho}'_{ge} = \boldsymbol{\varrho}^0_{gs}\,h_{se}\ ; \quad \boldsymbol{\sigma}'_{eg} = h_{es}\,\boldsymbol{\sigma}^0_{sg}\ ,$$

using the abbreviations

$$h_{se} = \left\langle s\left|\frac{\partial\mathscr{H}}{\partial Q}\right|e\right\rangle_0 (E_e - E_s)^{-1}$$

and

$$h_{es} = \left\langle e\left|\frac{\partial\mathscr{H}}{\partial Q}\right|s\right\rangle_0 (E_e - E_s)^{-1}$$

Here it is assumed that the state $|e\rangle$ of interest is coupled to only one other excited state $|s\rangle$. Thus Eq. (24) becomes

$$(\boldsymbol{\alpha}_{\rho\sigma})_{0\to1} = \left[\frac{\boldsymbol{\varrho}^0_{gs}\,h_{se}\,\boldsymbol{\sigma}^0_{eg}}{E_0 - E_L} + \frac{\boldsymbol{\varrho}^0_{ge}\,h_{es}\,\boldsymbol{\sigma}^0_{sg}}{E_1 - E_L}\right]\langle 1|Q|0\rangle \tag{25}$$

where $E_v = E_{e0} - E_{g0} + v\hbar\omega + i\,\Gamma_{ev}$ and the overlaps are omitted since $\langle 0|0\rangle = \langle 1|1\rangle = 1$. This is the basic equation of HT scattering for non-totally symmetric fundamentals. It clearly shows two resonances. In the first term the HT coupling is active in the emission process, supplying the borrowed transition moment, $\boldsymbol{\varrho}^0_{gs}$, and the energy denominator gives a resonance with the 0–0 absorption transition. In the second term the absorption moment, $\boldsymbol{\sigma}^0_{sg}$, is borrowed and resonance is therefore with the 0–1 vibronic sideband. The scattering processes for the 0–1 and 0–0 resonances are illustrated in Figs. 5, A and B respectively. Provided the bandwidths Γ_{ev} are not too large, the two resonances give rise to two peaks in the excitation profile, sometimes referred to as a Mortensen doublet *(41)*. This behaviour is observed generally in the porphyrin systems where the scattering for each non-totally symmetric mode reaches a maximum at the wavenumber of the Q_0 band (ν_{0-0}) and also at a frequency in the Q_1 region given approximately by $\nu_{0-0} + \nu_a$, ν_a being the wavenumber of the fundamental in each case *(42)*. For the case of coupling between two doubly-degenerate states (as in the porphyrins), $|\boldsymbol{\varrho}^0_{gs}| = |\boldsymbol{\sigma}^0_{sg}|$ and $|\boldsymbol{\varrho}^0_{ge}| = |\boldsymbol{\sigma}^0_{eg}|$. Therefore Eq. (25) predicts equal magnitudes for the two peaks of the excitation profile, since $|h_{es}| = |h_{se}|$ and $\Gamma_{e0} \approx \Gamma_{e1}$. The same is true in the weak coupling limit for totally symmetric modes when, in sufficiently high symmetries, $\rho = \sigma$ only.

Deviations from equal scattering intensity for the Q_0 and Q_1 bands have been observed in various metallo-porphyrins *(40)* and the results have been interpreted in terms of a combination of non-adiabatic coupling between the Q and B states and a Jahn-Teller distortion in the Q state *(40)*. The inverse-polarised bands arising from modes of A_2 symmetry, which are not Jahn-Teller active, show only the non-adiabatic effect which favours the Q_1 band scattering *(43)*. Non-adiabatic contributions to the scattering process are depicted in Fig. 6. This contrasts with the behaviour of the polarised and depolarised bands, most of which show stronger Q_0 scattering. The latter is thought to arise from additional FC A-term contributions resulting from Jahn-Teller distortions along the vibrational coordinates corresponding to the depolarised bands.

Enhanced scattering in the Q_1 band may also result from interference between the FC scattering of totally symmetric modes with shifted potentials and HT scattering by the non-totally symmetric mode concerned. *(44)*

2.9.1 Overtones and Combination Bands of Herzberg-Teller Active Modes

The absorption intensity of the 0–1 vibronic sideband is entirely borrowed by HT coupling. The observation that this 0–1 transition is of comparable intensity to the 0–0 transition suggests the feasibility of resonance Raman processes in which both the absorption and emission moments are borrowed. This second-order HT coupling displays itself in the appearance of first overtones and combination bands of non-totally symmetric modes. These are observed extensively in the haem proteins *(35)*.

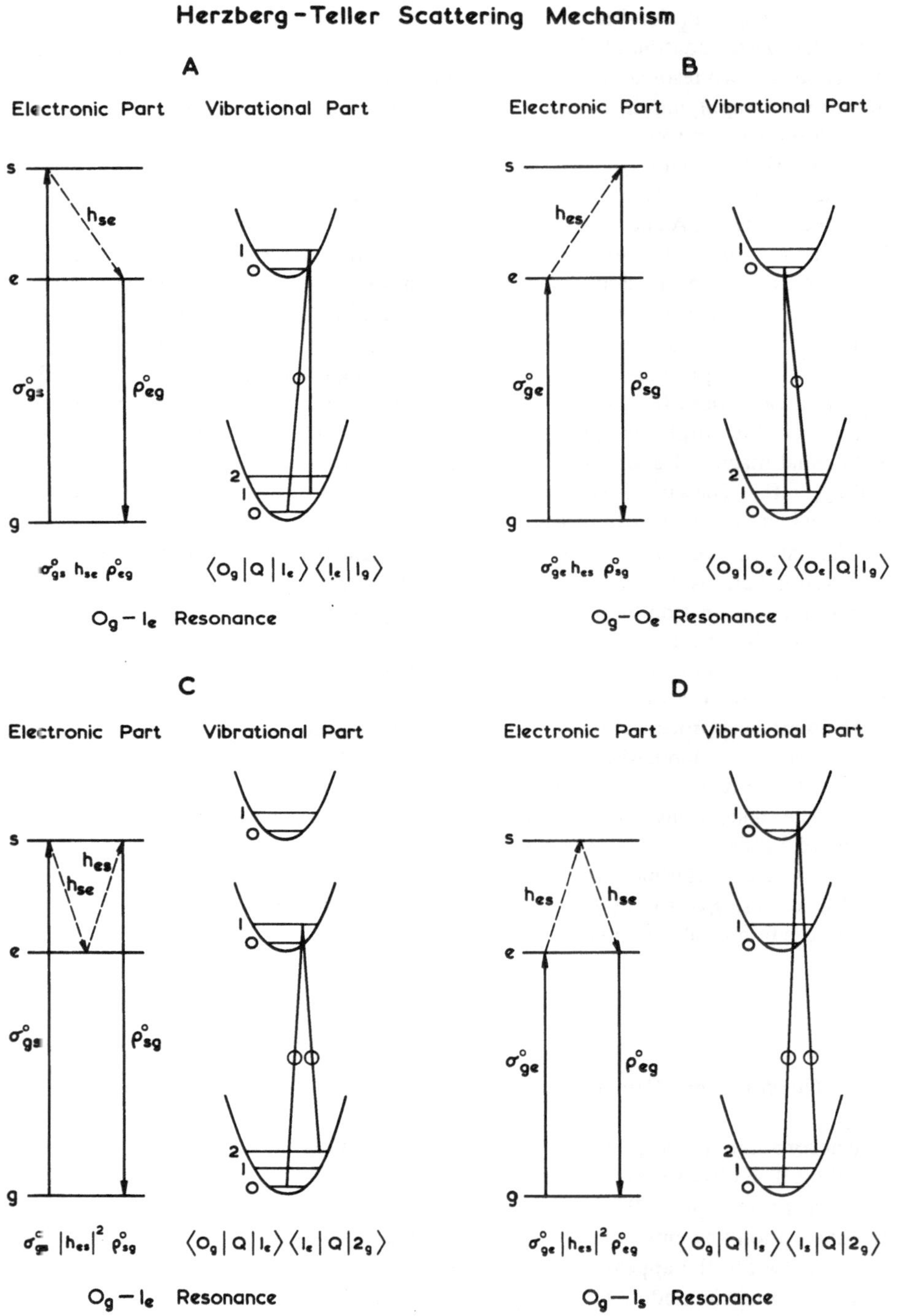
Herzberg - Teller Scattering Mechanism
A
Electronic Part
Vibrational Part
s
e
g
h_se
σ°_gs
ρ°_eg
σ°_gs h_se ρ°_eg
⟨0_g | Q | 1_e⟩ ⟨1_e | 1_g⟩
0_g − 1_e Resonance
B
Electronic Part
Vibrational Part
h_es
σ°_ge
ρ°_sg
σ°_ge h_es ρ°_sg
⟨0_g | 0_e⟩ ⟨0_e | Q | 1_g⟩
0_g − 0_e Resonance
C
Electronic Part
Vibrational Part
h_es
h_se
σ°_gs
ρ°_sg
σ°_gs |h_es|² ρ°_sg
⟨0_g | Q | 1_e⟩ ⟨1_e | Q | 2_g⟩
0_g − 1_e Resonance
D
Electronic Part
Vibrational Part
h_es
h_se
σ°_ge
ρ°_eg
σ°_ge |h_es|² ρ°_eg
⟨0_g | Q | 1_s⟩ ⟨1_s | Q | 2_g⟩
0_g − 1_s Resonance

◁

Fig. 5. The Herzberg-Teller scattering mechanism. A and B illustrate the first-order contributions to fundamental scattering for non-totally symmetric modes where the excited-state potential curves are unshifted relative to those of the ground state. C and D show the dominant second-order contributions to overtone scattering.
In the vibrational parts of the diagrams, vertical lines between potential curves represent overlap integrals, which are diagonal in the vibrational quantum number for unshifted harmonic potentials of the same fundamental frequency. An inclined line with a circle represents a matrix element of the coordinate operator, Q, which changes the vibrational quantum number by ± 1.
In the electronic parts of the diagrams, vertical arrows represent electric-dipole transitions effected by the transition-moment operators σ^0, ρ^0. A dashed arrow represents the vibronic coupling interaction between the two electronic excited states. $h_{se} = \langle s|\partial\mathscr{H}/\partial Q|e\rangle_0 (E_e-E_s)^{-1}$ and $\sigma^0_{gs} = \langle g|\sigma^0|s\rangle$, etc. In A, the HT interaction occurs in absorption giving a 0_g-1_e resonance. In B the HT interaction occurs in emission and a 0_g-0_e resonance results. C shows the production of an overtone ($0_g \rightarrow 2_g$) by HT coupling in absorption and emission. This second-order process displays only a 0_g-1_e resonance. D shows the second-order coupling for resonance with the $|s\rangle$ state. The second-order contributions to the Rayleigh scattering, from $\langle 0_g|Q|1_e\rangle\langle 1_e|Q|0_g\rangle$, are not shown

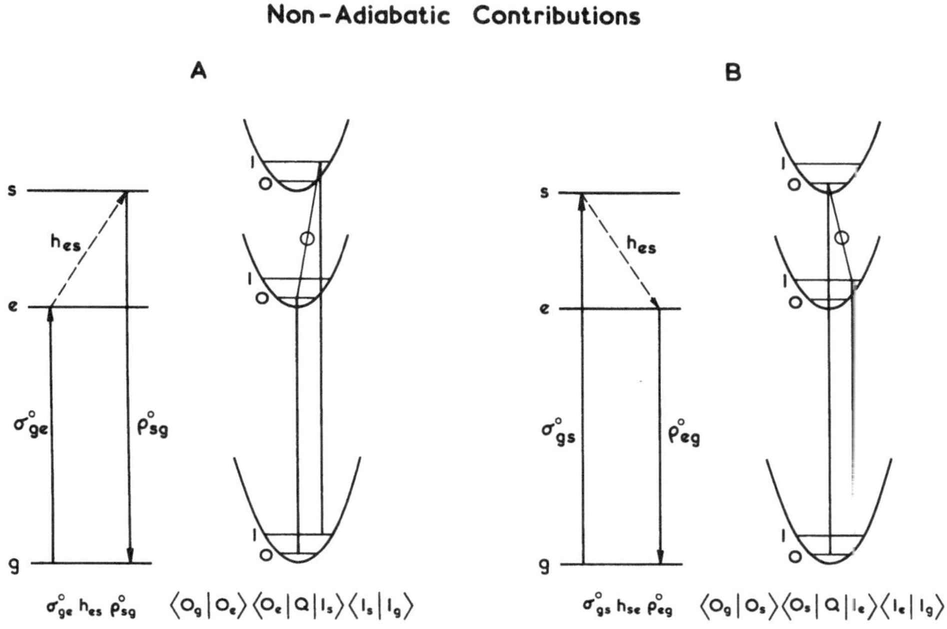

Fig. 6. Non-adiabatic contributions to the scattering process are significant when the electronic energy separation of excited states $|e\rangle$ and $|s\rangle$ is comparable to the vibrational energylevel separation. In this case the coordinate operator, Q, may cause transitions between the vibrational manifolds of the two electronic excited states

No overtones or combinations involving more than two quanta are observed, which is consistent with the absence of a 0–2 transition in absorption. The second-order bands in the resonance Raman spectrum of cytochrome c are shown in Fig. 20 (Section 4.8) where they appear in the 1650–3300 cm^{-1} region for excitation at 514.5 nm (point F). The excitation profiles of the second-order bands should peak only in the 0–1 absorption region, at positions consistent with the wavenumbers of the relevant fundamentals. This behaviour is expected since the appropriate parts of the scattering tensor show only 0–1 resonance conditions:

$$(\alpha_{\rho\sigma})_{0_a \rightarrow 2_a} = \varrho^0_{gs} |h_{es}(Q_a)|^2 \sigma^0_{sg} \langle 2|Q_a|1\rangle (E_1 - E_L)^{-1} \langle 1|Q_a|0\rangle \quad (26)$$

$$(\alpha_{\rho\sigma})_{0_a 0_b \rightarrow 1_a 1_b} = \varrho^0_{gs} h_{se}(Q_a) h_{es}(Q_b) \sigma^0_{sg} [\langle 1_a 1_b|Q_b|1_a 0_b\rangle (E_0 + \hbar\omega_a - E_L)^{-1} \langle 1_a 0_b|Q_a|0_a 0_b\rangle + \langle 1_a 1_b|Q_a|0_a 1_b\rangle (E_0 + \hbar\omega_b - E_L)^{-1} \langle 0_a 1_b|Q_b|0_a 0_b\rangle] \quad (27)$$

Indeed, for cytochrome c, the second-order bands appear to be absent for 0–0 (α band) excitation and show up only in the 0–1 (β band) resonance spectrum (*45*). Equation 26 shows the dominant contribution to an overtone $2\nu_a$, the scattering process being illustrated in Fig. 5C for resonance with the (lower energy) $|e\rangle$ state. For completeness, Fig. 5D shows the similar process for resonance with the upper, $|s\rangle$, state where both transition moments are borrowed from the lower $\langle g| \leftrightarrow |e\rangle$ transition. Contributions from matrix elements of the type $\langle 2|Q_a^2|0\rangle\langle 0|0\rangle$ and $\langle 2|2\rangle\langle 2|Q_a^2|0\rangle$ would give a 0–0 and a 0–2 resonance respectively but are likely to be insignificant since the second term also produces a 0–2 absorption which is generally not observed.

Equation (27) shows the contributions to the scattering tensor for a second-order combination band $\nu_a + \nu_b$. It should be noted that the vibrational coordinate operators act independently on their respective vibrational functions. The expression shows resonances with both $0-1_a$ and $0-1_b$ transitions so that the observation of resonance in an excitation profile is likely to be favoured if $\hbar\omega_a \approx \hbar\omega_b$. Again, the significance of terms of the type $\langle 1_a 1_b|Q_a Q_b|0_a 0_b\rangle\langle 0_a 0_b|0_a 0_b\rangle$ is likely to be contingent upon the appearance of the combination $0-1_a 1_b$ in the absorption spectrum.

2.9.2 Antisymmetric Tensor Contributions to Resonance Raman Scattering

In the normal vibrational Raman effect it has been traditionally assumed that the scattering tensor is symmetric ($\alpha_{\rho\sigma} = \alpha_{\sigma\rho}$). However, even in 1934 *Placzek* (*5*) considered the consequences of antisymmetric contributions, which he termed magnetic-dipole scattering because of the agreement between the selection rules for an antisymmetric Raman process and for a magnetic-dipole transition. Placzek gave the expected value of the depolarisation ratio, $\rho_1 = I_\perp/I_\parallel = \infty$ for purely antisymmetric scattering, and also considered the scattering of circularly polarised light. The behaviour of ρ_1, and the symmetry aspects of antisymmetric scattering, are discussed in Section 2.11.

The occurrence of antisymmetric scattering contributions is invariably associated with processes in which the symmetry of particular electronic states becomes important. In the non-resonance electronic (or the vibro-electronic) Raman effect (ERE or VERE), the initial and final states are different electronic (or vibronic) states. Thus, if these states are of appropriate symmetry, an antisymmetric contribution may be present in the scattering (*46*). The same is true for the special case of the VERE represented by vibrational Raman scattering in systems possessing degenerate electronic ground states (see Section 2.11). In the resonance ERE and resonance VERE as well as the vibrational resonance-Raman effect, it is the symmetry of the intermediate state which becomes significant.

Two important examples in the vibrational resonance-Raman effect will be described in detail here. in *Case A*, a purely antisymmetric vibrational mode is considered. Such a mode is inactive off resonance but may give rise to resonance scattering since it can be active in vibronic coupling in a degenerate excited state, or between two distinct degenerate excited states. A notable example of this phenomenon is furnished by the resonance Raman spectrum of oxyhaemoglobin (Section 4.8) which contains several inverse polarised bands ($I_{\parallel} = 0, I_{\perp} \neq 0, \rho_1 = \infty$). *Case B* is that of a mode for which the scattering tensor has mixed symmetric and antisymmetric parts, i.e. $\boldsymbol{\alpha}$ is asymmetric ($\boldsymbol{\alpha}_{\rho\sigma} \neq \boldsymbol{\alpha}_{\sigma\rho}$). For such a mode, only the symmetric part is active off resonance. In the simplest case, a non-degenerate asymmetric mode vibronically couples two non-degenerate transitions with orthogonal polarisation, for example an $A_2(R_z, xy)$ mode in C_{2v} coupling $B_1(x)$ and $B_2(y)$ transitions. These modes exhibit anomalous polarisation in the resonance region ($\frac{3}{4} < \rho_1 < \infty$), as exemplified by several bands of cytochrome c (see Section 4.8).

Case A. Consider the HT coupling between two doubly-degenerate excited states, $|e\rangle$ and $|s\rangle$, as for example between two distinct E_u states in a molecule of D_{4h} symmetry. Transitions to each state are allowed, from an A_{1g} ground state, with both x and y polarisation. In addition to the zero-order transition moments $\mathrm{x}^0, \mathrm{y}^0$, the first-order borrowed moments x', y', must be considered. In order to demonstrate the origin of the antisymmetry of the tensor in this case ($\boldsymbol{\alpha}_{xy} = -\boldsymbol{\alpha}_{yx}$) the non-zero off-diagonal elements are written as follows:

$$\left.\begin{aligned}
\boldsymbol{\alpha}_{xy} &= \left[\frac{\mathrm{x}'\mathrm{y}^0}{E_0 - E_L} + \frac{\mathrm{x}^0\mathrm{y}'}{E_1 - E_L} + \left(\frac{\mathrm{y}'\mathrm{x}^0}{E_0 - \hbar\omega + E_L} + \frac{\mathrm{y}^0\mathrm{x}'}{E_1 - \hbar\omega + E_L}\right)\right] \langle 0|Q|1\rangle \\
\boldsymbol{\alpha}_{yx} &= \left[\frac{\mathrm{y}'\mathrm{x}^0}{E_0 - E_L} + \frac{\mathrm{y}^0\mathrm{x}'}{E_1 - E_L} + \left(\frac{\mathrm{x}'\mathrm{y}^0}{E_0 - \hbar\omega + E_L} + \frac{\mathrm{x}^0\mathrm{y}'}{E_1 - \hbar\omega + E_L}\right)\right] \langle 0|Q|1\rangle
\end{aligned}\right\} \quad (28)$$

where the non-resonant parts are in round brackets. To demonstrate the antisymmetry of the tensor in the case of resonance with the state $|e\rangle$ (components $|e,x\rangle$, $|e,y\rangle$) it is sufficient to show that and

$$\left.\begin{aligned} x'_{ge}\, y^0_{eg} &= -\, y'_{ge}\, x^0_{eg} \\ x^0_{ge}\, y'_{eg} &= -\, y^0_{ge}\, x'_{eg} \end{aligned}\right\} \tag{29}$$

Since $|e\rangle$ is doubly degenerate, $x^0_{eg} = y^0_{eg}$ and $x^0_{ge} = y^0_{ge}$. Thus Eqs. (29) become

$$x'_{ge} = -\, y'_{ge}$$

Introducing the HT-borrowed moments from the $|s\rangle$ state

$$\begin{aligned} x'_{ge} &= x^0_{gs}\, \langle s, x \left| \frac{\partial \mathcal{H}}{\partial Q} \right| e, \beta\rangle\, (E_e - E_s)^{-1} \\ y'_{ge} &= y^0_{gs}\, \langle s, y \left| \frac{\partial \mathcal{H}}{\partial Q} \right| e, \beta\rangle\, (E_e - E_s)^{-1} \end{aligned} \tag{30}$$

Whether $\beta = x$ or y for the $|e, \beta\rangle$ state is determined by the symmetry of the vibrational co-ordinate Q. The Wigner-Eckart theorem allows the component-dependence of the vibronic-coupling matrix elements to be expressed, using the V coefficients of *Griffith* (*47*), by

$$\langle s, \alpha \left| \frac{\partial \mathcal{H}}{\partial Q} \right| e, \beta\rangle = V\begin{pmatrix} E_u & E_u & \Gamma_a \\ \alpha & \beta & \gamma \end{pmatrix} \langle s \left\| \frac{\partial \mathcal{H}}{\partial Q} \right\| e\rangle \tag{31}$$

where $\langle s \left\| \frac{\partial \mathcal{H}}{\partial Q} \right\| e\rangle$ is a reduced matrix element which is independent of the components α, β, γ. Now if $\Gamma_Q = A_{2g} (\gamma = \iota)$, the pure antisymmetric representation in D_{4h}, then the V coefficient is zero for $\alpha = \beta$ and

$$V\begin{pmatrix} E_u & E_u & A_{2g} \\ x & y & \iota \end{pmatrix} = -\, V\begin{pmatrix} E_u & E_u & A_{2g} \\ y & x & \iota \end{pmatrix}$$

This establishes the antisymmetry of α through Eq. (30), since again $x^0_{gs} = y^0_{gs}$ because of the (x, y)-degeneracy of $|s\rangle$. The result still holds if e = s.

As a further example, the V coefficient relevant to antisymmetric scattering in cubic symmetries is (*48*)

$$V\begin{pmatrix} T_1 & T_1 & T_1 \\ \alpha & \beta & \gamma \end{pmatrix} = V\begin{pmatrix} T_2 & T_2 & T_1 \\ \alpha & \beta & \gamma \end{pmatrix} = -\frac{\varepsilon_{\alpha\beta\gamma}}{\sqrt{6}} \tag{32}$$

where the alternating tensor $\varepsilon_{\alpha\beta\gamma}$ is antisymmetric in any pair of indices and is zero if any of α, β, γ are equal. The first V coefficient in Eq. (32) may arise from, for example, vibronic coupling via a T_{1g} mode between two T_{1u} excited states in O_h symmetry. The second V coefficient is appropriate to the coupling between two T_2 states via a T_1 mode in T_d symmetry.

The description outlined above using the V coefficients is generally applicable in any symmetry. An explicit demonstration of the antisymmetry of the scattering tensor for an A_{2g} vibrational mode in D_{4h} symmetry may be given following the treatment of *Warshel* (*49*). Such a pure antisymmetric vibrational mode is illustrated schematically in Fig. 7A. This mode causes a rotation of the transition moments in the x,y-plane whilst preserving their degeneracy. Figure 7B illustrates the variation in the transition moments during an A_{2g} vibration. The variations in the projection of x on x^0 and of y on y^0 are of the same sign but both these variations vanish in the limit $Q \to 0$. This corresponds to the expected absence of diagonal tensor elements for an A_{2g} mode in the equilibrium D_{4h} symmetry. [$\alpha \neq \beta$ in Eq. (31)]. In the limit $Q \to 0$, the only non-vanishing variations are in the off-diagonal projections shown by the white arrows of Fig. 7B. These variations, x′ and y′ are seen to be opposite in sign, thus demonstrating the required antisymmetry relation: $x^0 y' = - y^0 x'$.

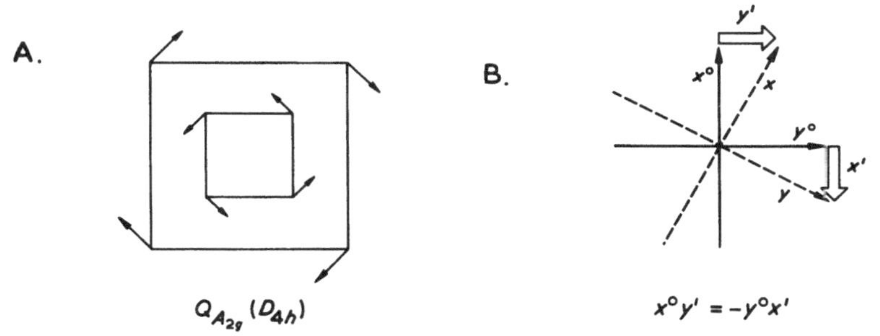

Fig. 7. A. Schematic illustration of an antisymmetric vibrational mode of A_{2g} symmetry in the D_{4h} point group. The diagram can represent a much simplified view of a porphyrin-type molecule. B. Illustration of the variation in the transition moments of E_u (D_{4h}) symmetry during an A_{2g} vibration. The full arrows represent the transition moments x^0, y^0 corresponding to the equilibrium configuration $Q_{A_{2g}} = 0$. The dashed arrows are the moments x, y corresponding to a turning point of the vibration. In the limit $Q \to 0$, the only non-vanishing variations are shown by the white arrows

Returning now to Eqs. (28), it can be seen that, because of the antisymmetry relations, the resonant and non-resonant parts of each tensor element tend to cancel in the off-resonance limit: $E_0 \approx E_1 \gg E_L \gg \hbar\omega$. In addition, the contributions from $|e\rangle$ and $|s\rangle$ states to each term in Eqs. (28) are of comparable magnitude off resonance. These contributions tend to cancel because of the difference in sign of the HT perturbation energy denominators, viz. $(E_e - E_s)^{-1} = -(E_s - E_e)^{-1}$. The net result is the disappearance of antisymmetric scattering in the off-resonance region.

Case B. In this case $|e\rangle$ and $|s\rangle$ are non-degenerate. Let transitions $|g\rangle \rightarrow |e\rangle$ and $|g\rangle \rightarrow |s\rangle$ be allowed with x and y polarisation respectively. Thus $y^0_{ge} = y^0_{eg} = 0$ and $x^0_{gs} = x^0_{sg} = 0$ and, for resonance with the $|g\rangle \rightarrow |e\rangle$ transition, Eqs. (28) reduce to

$$\begin{aligned} \alpha_{xy} &= \left[\frac{x^0_{ge}\, y'_{eg}}{E_1 - E_L}\left(+\frac{y'_{ge}\, x^0_{eg}}{E_0 - \hbar\omega + E_L}\right)\right]\langle 0|Q|1\rangle \\ \alpha_{yx} &= \left[\frac{y'_{ge}\, x^0_{eg}}{E_0 - E_L}\left(+\frac{x^0_{ge}\, y'_{eg}}{E_1 - \hbar\omega + E_L}\right)\right]\langle 0|Q|1\rangle \end{aligned} \qquad (33)$$

So, in this case $\alpha_{xy} \neq \alpha_{yx}$ owing only to the difference in the energy denominators. The tensor is therefore asymmetric. In resonance with the 0–0 transition, scattering is dominated by the element α_{yx} and, for the 0–1 resonance, by the element α_{xy}. The behaviour of the depolarisation ratio in this case (anomalous polarisation) is discussed in Section 2.11.2.

2.10 Combination-Band Progressions Involving One or Two Quanta of a Non-Totally Symmetric Mode

The purpose of this section is to draw attention briefly to two interesting examples of progressions involving both totally symmetric and non-totally symmetric modes. First, the resonance Raman spectra (*50*) of various salts of the $[AuBr_4]^-$ ion (D_{4h} symmetry) show progressions in the ν_1 (A_{1g}) Au–Br stretching frequency based on one quantum of ν_2 (B_{1g}) (up to $\nu_2 + 2\nu_1$) and also based on a quantum of ν_4 (B_{2g}) (up to $\nu_4 + 5\nu_1$). Excitation was at 457.9 nm, which lies near the absorption maximum (22,500 cm^{-1}) of a broad band assigned (*51*) to a superposition of two electric-dipole-allowed transitions, both of the type $\sigma^* d_{x^2-y^2}(Au) \leftarrow \pi^b p(Br)$, ${}^1A_{2u} \leftarrow {}^1A_{1g}(b_{1g} \leftarrow b_{2u})$ and ${}^1E_u \leftarrow {}^1A_{1g}(b_{1g} \leftarrow e_u)$. Despite the favourable proximity of the two excited states, no HT coupling between them is possible owing to the absence of the required modes of E_g symmetry in the $[AuBr_4]^-$ ion. It seems likely that the origin of the resonance Raman progressions involving the B_{1g} and B_{2g} modes is a Jahn-Teller effect in the excited ${}^1E_u$ state (*52*) since the symmetries of these modes comprise the symmetric part of the direct product $E_u \otimes E_u$. It is unlikely that an explanation could involve a lowering of symmetry for the ground state in the solid since the B_{1g} and B_{2g} mode fundamentals display preresonance enhancement in solution where the symmetry of the ion must surely be D_{4h}.

The second case of interest concerns the appearance of even numbers of quanta of IR-active (Raman-inactive) modes as origins for progressions in totally symmetric modes. This occurs in the resonance Raman spectra of the ions $[Ru_2OCl_{10}]^{4-}$ and $[Re_2OCl_{10}]^{3-}$ where, most notably, long progressions in the ν_1 (A_{1g}) M–O–M symmetric stretching frequency are based on origins of $2\nu_9$ and $4\nu_9$. (See Table 10 in Section 4.6 for a full list of observed progressions). ν_9 has been assigned as the IR-active bending fundamental of E_u symmetry. In the case of $[Ru_2OCl_{10}]^{4-}$, a simple MO treatment predicts three possible electronic transitions in the region of resonance excitation. The transitions, which are all to the π^*(Ru–O) orbital localised largely on the Ru atoms, originate from the three non-bonding combinations of Ru d orbitals, viz.

$$
\begin{array}{lll}
{}^1A_{1u}, {}^1A_{2u}, {}^1B_{1u}, {}^1B_{2u} \leftarrow {}^1A_{1g} & & (e_u^* \leftarrow e_g) \\
{}^1E_u \leftarrow {}^1A_{1g} & & (e_u^* \leftarrow b_{2g}) \\
{}^1E_g \leftarrow {}^1A_{1g} & & (e_u^* \leftarrow b_{1u})
\end{array}
$$

The first transition is allowed with z-polarisation through the ${}^1A_{2u}$ component, the second transition is allowed with x,y-polarisation and the third is Laporte-forbidden. The resonant electronic transition has been assigned as the ${}^1A_{2u} \leftarrow {}^1A_{1g}$ on the grounds of the observed value of the depolarisation ratio ($\rho_l = 1/3$, see Section 2.11.1) for the totally symmetric Raman bands at resonance. However, the appearance of even quanta of the E_u mode is remarkable. The totally symmetric part of these even overtones can only be Franck-Condon active if there is a large difference in fundamental frequency for the E_u mode between the ground and excited states. The latter may result from strong HT coupling of the resonant excited state with higher energy states. The appearance of the E_u modes is thus strongly indicative of HT coupling between the ${}^1A_{2u}$ and ${}^1E_g$ states. All other HT couplings amongst the three transitions must involve E_g modes which, however, have not been identified in the resonance Raman progressions. The above hypothesis of HT coupling via the E_u mode is equally consistent with a resonance with the ${}^1E_g \leftarrow {}^1A_{1g}$ Laporte-forbidden transition. In this case the even quanta are associated with the borrowing of both absorption and emission moments by a second-order HT mechanism (see Section 2.9.1). The borrowed moments are both z-polarised and thus only $\boldsymbol{\alpha}_{zz} \neq 0$, giving $\rho_l = 1/3$ for this resonance also. The appearance of $4\,\nu_9$ in the resonance Raman spectrum, via $\langle 0|Q^2|2\rangle\langle 2|Q^2|4\rangle$ in this mechanism, would imply the presence of $2\,\nu_9$ in the absorption spectrum which, however, is unresolved. The problem seems also to be unresolved but it is clear that interactions involving the Laporte-forbidden transition cannot be neglected in this case.

2.11 Depolarisation Ratios and the Symmetry of the Scattering Tensor

The definition of the depolarisation ratio, ρ_l, is illustrated in Fig. 1 for linearly-polarised incident radiation and a 90° scattering geometry. In the normal Raman effect it is well known that the measurement of ρ_l may identify the symmetry of the vibrational mode responsible for a given Raman band; $\rho_l < 3/4$ ($\rho_l = 0$ in cubic or higher symmetries) for totally symmetric modes and $\rho_l = 3/4$ for non-totally symmetric modes. In the resonance Raman effect, the value of ρ_l, and its dependence on the exciting frequency, may be more informative. This is because the symmetries of particular electronic states, as well as the symmetry aspects of their vibronic interactions become significant in resonance.

The quantities relevant to the rotationally averaged situation of randomly oriented species in solution or the gas phase must necessarily be invariants of the rotational symmetry. Accordingly, they must transform under the irreducible representations of the rotation group in three dimensions (without inversion), R_3, just like the angular momentum functions of an atom. The polarisability, $\boldsymbol{\alpha}_{\rho\sigma}$, is a second-rank cartesian tensor and gives rise to three irreducible tensors (*53*), $\boldsymbol{\alpha}^{(0)}$, $\boldsymbol{\alpha}^{(1)}$, $\boldsymbol{\alpha}^{(2)}$, corresponding in rotational behaviour to the spherical harmonics, Y^{l}_{m}, with $l = 0, 1, 2$ respectively. The components $\boldsymbol{\alpha}^{(l)}_{m}$, $-l \leqslant m \leqslant l$, of the irreducible tensors are given below.

$$\boldsymbol{\alpha}^{(0)}_{0} = -3^{-1/2}\,(\boldsymbol{\alpha}_{xx} + \boldsymbol{\alpha}_{yy} + \boldsymbol{\alpha}_{zz})$$

$$\left\{\begin{aligned} \boldsymbol{\alpha}^{(1)}_{1} &= \tfrac{1}{2}\,(-\boldsymbol{\alpha}_{xz} + \boldsymbol{\alpha}_{zx} - \mathrm{i}\boldsymbol{\alpha}_{yz} + \mathrm{i}\boldsymbol{\alpha}_{zy}) \\ \boldsymbol{\alpha}^{(1)}_{0} &= 2^{-1/2}\,(\mathrm{i}\boldsymbol{\alpha}_{xy} - \mathrm{i}\boldsymbol{\alpha}_{yx}) \\ \boldsymbol{\alpha}^{(1)}_{-1} &= -\tfrac{1}{2}\,(\boldsymbol{\alpha}_{xz} - \boldsymbol{\alpha}_{zx} - \mathrm{i}\boldsymbol{\alpha}_{yz} + \mathrm{i}\boldsymbol{\alpha}_{zy}) \end{aligned}\right.$$

$$\left\{\begin{aligned} \boldsymbol{\alpha}^{(2)}_{2} &= \tfrac{1}{2}\,(\boldsymbol{\alpha}_{xx} - \boldsymbol{\alpha}_{yy} + \mathrm{i}\boldsymbol{\alpha}_{xy} + \mathrm{i}\boldsymbol{\alpha}_{yx}) \\ \boldsymbol{\alpha}^{(2)}_{1} &= \tfrac{1}{2}\,(\boldsymbol{\alpha}_{xz} + \boldsymbol{\alpha}_{zx} + \mathrm{i}\boldsymbol{\alpha}_{yz} + \mathrm{i}\boldsymbol{\alpha}_{zy}) \\ \boldsymbol{\alpha}^{(2)}_{0} &= 6^{-1/2}\,(2\boldsymbol{\alpha}_{zz} - \boldsymbol{\alpha}_{xx} - \boldsymbol{\alpha}_{yy}) \\ \boldsymbol{\alpha}^{(2)}_{-1} &= \tfrac{1}{2}\,(\boldsymbol{\alpha}_{xz} + \boldsymbol{\alpha}_{zx} - \mathrm{i}\boldsymbol{\alpha}_{yz} - \mathrm{i}\boldsymbol{\alpha}_{zy}) \\ \boldsymbol{\alpha}^{(2)}_{-2} &= \tfrac{1}{2}\,(\boldsymbol{\alpha}_{xx} - \boldsymbol{\alpha}_{yy} - \mathrm{i}\boldsymbol{\alpha}_{xy} - \mathrm{i}\boldsymbol{\alpha}_{yx}) \end{aligned}\right.$$

In the molecular point groups, the three trace elements of $\boldsymbol{\alpha}^{(0)}_{0}$ always transform under the totally symmetric representation. The symmetry behaviour of the three components of $\boldsymbol{\alpha}^{(1)}$ corresponds with that of the components of the magnetic dipole operator, which transform like the rotations R_x, R_y, R_z. The components of $\boldsymbol{\alpha}^{(2)}$ transform like those of the quadrupole moment operator, that is, like the five d orbitals.

These quantities give rise to the three basic types of contribution to the scattering expressed by the three invariants [1]):

[1]) The definitions of the invariants used here follow those of *Placzek* (*5d*) and are preferred because of their simple relation to the irreducible tensors. Many authors use the invariants $\bar{\alpha}^2$, γ_s^2, γ_a^2 where $\bar{\alpha}^2 = \frac{1}{3}G^0$, $\gamma_s^2 = \frac{3}{2}G^s$ and $\gamma_a^2 = \frac{3}{2}G^a$. Values of the depolarisation ratio are unaffected but the constant of proportionality in intensity expressions will be different. The quantities β_c, β_a and γ_s^2 of Ref. (*54*) correspond exactly with G^0, G^a and G^s respectively.

a) the isotropy; $G^0 = |\boldsymbol{\alpha}_0^{(0)}|^2$ ($= 3\bar{\alpha}^2$ where $\bar{\alpha}$ is the mean polarisability).

b) the symmetric anisotropy; $G^s = \sum_m |\boldsymbol{\alpha}_m^{(2)}|^2$;

c) the antisymmetric anisotropy; $G^a = \sum_m |\boldsymbol{\alpha}_m^{(1)}|^2$.

To assist in the evaluation of the invariants in terms of elements of the cartesian polarisability tensor, it is convenient to define the symmetric and antisymmetric tensors: $S_{\rho\sigma} = \frac{1}{2}(\boldsymbol{\alpha}_{\rho\sigma} + \boldsymbol{\alpha}_{\sigma\rho})$; $A_{\rho\sigma} = \frac{1}{2}(\boldsymbol{\alpha}_{\rho\sigma} - \alpha_{\sigma\rho})$. Now

$$\begin{aligned} G^0 &= \tfrac{1}{3}\,|\mathrm{Tr}\{S\}|^2 \\ G^s &= \mathrm{Tr}\{S\}\{S^\dagger\} - G^0 \\ G^a &= \mathrm{Tr}\{A\}\{A^\dagger\} \end{aligned} \qquad (34)$$

where Tr indicates the trace and † denotes the transpose of the tensor (or the Hermitian conjugate if $\boldsymbol{\alpha}$ is complex).

The scattered intensity may be expressed in terms of these invariants by a transformation from the molecule-fixed coordinates of the molecular polarisability to the laboratory-fixed coordinates (Fig. 1) appropriate to the measurement (*54*). Thus

$$I_{\mathrm{TOTAL}}(= I_{\parallel} + I_{\perp}) \propto 10\,G^0 + 7\,G^s + 5\,G^a$$

and the scattered intensities parallel and perpendicular to the incident polarisation are respectively

$$I_{\parallel} \propto 10G^0 + 4\,G^s$$

and

$$I_{\perp} \propto 3\,G^s + 5\,G^a$$

for a 90° scattering geometry. Hence,

$$\rho_l = \frac{I_{\perp}}{I_{\parallel}} = \frac{3\,G^s + 5\,G^a}{10\,G^0 + 4\,G^s} \qquad (35)$$

Thus, when each of the invariants contributes alone, a characteristic value of ρ_l results:

(a) Pure isotropic scattering: $G^0 \neq 0$, $G^s = 0$, $G^a = 0$
$\rho_l = 0$ (independent of the value of G^0)
(b) Pure symmetric scattering: $G^0 = 0$, $G^s \neq 0$, $G^a = 0$
$\rho_l = 3/4$ (independent of the value of G^s)
(c) Pure antisymmetric scattering: $G^0 = 0$, $G^s = 0$, $G^a \neq 0$
$\rho_l = \infty$ (independent of the value of G^a)

It must be noted, however, that in case (c), if $G^a = 0$ due to the symmetry of the scattering tensor ($\boldsymbol{\alpha}_{\rho\sigma} = \boldsymbol{\alpha}_{\sigma\rho}$), then $I_{\parallel} = I_{\perp} = 0$ and no scattering is observed. For each of these pure types of scattering the value of ρ_l is a constant, independent of the excitation energy. Thus dispersion of ρ_l with excitation frequency requires mixed contributions from the invariants. This occurs when the scattering species belongs to a point group in which two, or all three, of the invariants transform under the same irreducible representation.

The symmetries of the scattering tensor in the 32 molecular point groups have been given by *McClain* (*55*) and this allows the contributions from any of the invariants to be evaluated for a Raman process transforming under any irreducible representation. Experimentally, the evaluation of all three tensor invariants requires three independent intensity measurements. The measurement of the total scattering at 90°, ($I_{\parallel} + I_{\perp}$), and the depolarisation ratio, ρ_l, may be supplemented by a measurement involving circularly polarised light. It is usual to measure the reversal coefficient (*5c*), $P(\pi) = I_{contra}/I_{co}$, which is the ratio of the intensity of contra-rotating to that of co-rotating scattered light for the back scattering of pure circularly-polarised incident radiation. In terms of the invariants

$$P(\pi) = \frac{10\,G^0 + G^s + 5\,G^a}{6\,G^s}$$

For forward scattering, $P(0) = P(\pi)^{-1}$. Alternatively, the degree of circularity, C, may be measured.

$$C(\pi) = (I_{co} - I_{contra})/(I_{co} + I_{contra}) = \frac{10\,G^0 - 5\,G^s + 5\,G^a}{10\,G^0 + 7\,G^s + 5\,G^a}$$

The three measurements have been used to elucidate the contributions from each of the three invariants to the scattering intensity of various Raman bands in cytochrome c (*56*).

2.11.1 Totally Symmetric Modes

The values expected for ρ_l in a few cases of special interest will now be discussed. First, in cubic symmetries, inspection of character tables shows that the trace of the scattering tensor transforms (like $x^2 + y^2 + z^2$) alone under the totally symmetric irreducible representation. Thus a totally symmetric vibrational mode will display pure isotropic scattering in this case with $\rho_l = 0$. As mentioned previously, no dispersion of ρ_l is expected. Explicitly, this is because the allowed electric-dipole transition moments are triply degenerate and thus a resonance effect does not alter the equivalence of the trace elements. Any vibronic coupling contributions are also equivalent for the three different polarisation directions when a totally symmetric mode is involved.

Unless the symmetry is very low, only the diagonal tensor elements contribute to the scattering for totally symmetric modes, as in the example just discussed. However, in symmetries lower than cubic, the symmetry of the resonant electronic transition is significant (*57*). If the transition is non-degenerate, only a single diagonal tensor element (α_{zz} say) is dominant at resonance. Thus, from Eqs. (34), $G^0 = \frac{1}{3}\alpha_{zz}^2$, $G^s = \alpha_{zz}^2 - \frac{1}{3}\alpha_{zz}^2 = \frac{2}{3}\alpha_{zz}^2$, $G^a = 0$, and $\rho_l = 3\,G^s/(10\,G^0 + 4\,G^s) = \frac{1}{3}$ at resonance. If the transition is doubly degenerate then $\alpha_{zz} = 0$ (say) and $\alpha_{xx} = \alpha_{yy}$ at resonance. Now, $Tr\{S\} = 2\,\alpha_{xx}$ and so $G^0 = \frac{4}{3}\alpha_{xx}^2$, $G^s = 2\alpha_{xx}^2 - \frac{4}{3}\alpha_{xx}^2 = \frac{2}{3}\alpha_{xx}^2$, $G^a = 0$, resulting in $\rho_l = 1/8$ at resonance. The measurement of ρ_l at resonance for a totally symmetric mode can therefore indicate the assignment of the resonant electronic transition in certain cases (*58*), since the value off-resonance will not necessarily be 1/3 or 1/8 accordingly, but will depend on the particular value of the anisotropy. Thus, in the point groups lower than cubic, there is generally mixed isotropic and symmetric scattering. Additionally, in the groups C_1, C_s, C_i, C_n, C_{nh} and S_n, the antisymmetric anisotropy contributes to the scattering by totally symmetric modes, since R_z transforms under the totally symmetric representation. This contribution vanishes off resonance for the pure vibrational Raman effect since the tensor is symmetric ($\alpha_{\rho\sigma} = \alpha_{\sigma\rho}$) and thus $G^a = 0$. In resonance with a non-degenerate transition which is z-polarised, only $\alpha_{zz} \neq 0$ and so $\rho_l = 1/3$ as above. However, for resonance with an E-type transition, the tensor has the form

$$\begin{pmatrix} a & b & 0 \\ -b & a & 0 \\ 0 & 0 & 0 \end{pmatrix}$$

for totally symmetric modes. Thus $G^0 = \frac{4}{3}|a|^2$, $G^s = \frac{2}{3}|a|^2$, $G^a = 2|b|^2$ and $\rho_l = \frac{1}{8} + \frac{5}{8}\cdot\frac{|b|^2}{|a|^2}$. Anomalous polarisation ($\rho_l > 3/4$) may therefore be observed for $|\alpha_{xy}|^2 > |\alpha_{xx}|^2$.

If totally symmetric modes are excited as part of a vibroelectronic Raman effect then the symmetry of the electronic Raman transition determines the form of the tensor and, even off-resonance, antisymmetric contributions may be present (*59*). A special case of this type occurs in the "vibrational" resonance Raman of systems with degenerate electronic ground states (*60*). Here antisymmetric contributions may arise from Raman transitions between different degenerate components of the electronic ground state. Anomalous polarisation of the totally symmetric stretching fundamental and its overtones has been observed in the resonance Raman spectra of the $IrCl_6^{2-}$ (*61*) and $IrBr_6^{2-}$ (*62*) ions [ground states: $E_g''(O_h^*)\,{}^2T_{2g}(O_h)$]. The tensor patterns and expected values of ρ_l for various resonant intermediate states in $IrBr_6^{2-}$ have been given by *Hamaguchi* 1977 (*63*).

Table 1 shows the distribution of the contributions from the invariants of the scattering tensor for totally symmetric Raman processes in the molecular point groups.

Table 1. Totally symmetric Raman processes

			G^0	$G^0/G^s/G^a$	G^0/G^s
ρ_l	Off resonance		0	$0 < \rho_l < 3/4$	$0 < \rho_l < 3/4$
	On resonance	non-degenerate transition	–	$1/3 < \rho_l < 2$	1/3
		degenerate transition	0	$1/8 < \rho_l < 2$	1/8
Point Groups			$T\ T_d\ T_h$ $O\ O_h$ $I\ I_h$	$C_1\ C_s\ C_i$ $C_n\ C_{nh}\ S_n$	all groups except those in the other two columns

2.11.2 Non-Totally Symmetric Modes

For non-totally symmetric modes in any point group, there are no trace contributions. Therefore $G^0 = 0$ and ρ_l is given by the special formula

$$\rho_l = \frac{3}{4} + \frac{5}{4}\frac{|G^a|^2}{|G^s|^2} \tag{36}$$

which, using Eqs. (*34*), may also be written as

$$\rho_l = \frac{3}{4} + \frac{5}{4} \cdot \frac{|\boldsymbol{\alpha}_{\rho\sigma} - \boldsymbol{\alpha}_{\sigma\rho}|^2}{|\boldsymbol{\alpha}_{\rho\sigma} + \boldsymbol{\alpha}_{\sigma\rho}|^2}$$

Thus the value of ρ_l depends on the degree of asymmetry in the tensor. For pure symmetric scattering, as in the normal Raman effect for instance, $\boldsymbol{\alpha}_{\rho\sigma} = \boldsymbol{\alpha}_{\sigma\rho}$ and ρ_l = 3/4 (depolarised scattering). For pure antisymmetric scattering $\boldsymbol{\alpha}_{\rho\sigma} = -\boldsymbol{\alpha}_{\sigma\rho}$ and therefore $G^s = 0$, giving $\rho_l = \infty$ (inverse polarisation). For an asymmetric tensor $|\boldsymbol{\alpha}_{\rho\sigma}| \neq |\boldsymbol{\alpha}_{\sigma\rho}|$ and $\frac{3}{4} < \rho_l < \infty$, which is called anomalous polarisation.

In all point groups possessing two-fold (x,y) degenerate representations (except C_n, C_{nh} and S_n) there is one pure antisymmetric representation of the A_2 type. If a molecule, belonging to one of these point groups, possesses a vibrational mode of this symmetry, then the mode is inactive in the normal Raman effect due to the symmetry of the tensor. However, the mode may give rise to scattering for resonance with a double-degenerate electronic transition since it is $(\boldsymbol{\alpha}_{xy} - \boldsymbol{\alpha}_{yx})$ which transforms under the A_2-type irreducible representation and therefore the mode is only active in vibronic coupling between the x and y components of the electronic excited state (or of two different excited states). Since these modes are purely antisymmetric they appear with inverse polarisation ($\rho_l = \infty$) and show no dispersion in ρ_l.

Table 2. Non-totally symmetric Raman processes

		G^s	G^a	Mixed G^s/G^a
ρ_l	Off resonance	3/4	–	3/4
	On resonance	3/4	∞	$2\ (2 < \rho_l^{max} < \infty)$
Point Group				
C_s				A''
C_2				B
C_3				E
C_4		B		E
C_5				
C_6		} E_2		} E_1
C_7				
C_8				
D_2				B_1, B_2, B_3
D_3				E
D_4		B_1, B_2	} A_2	E
D_5		E_2		E_1
D_6		E_2		E_1
C_{2v}				A_2, B_1, B_2
C_{3v}				E
C_{4v}		B_1, B_2	} A_2	E
C_{5v}		E_2		E_1
C_{6v}		E_2		E_1
C_{2h}				B_g
C_{3h}		E'		E''
C_{4h}		B_g		E_g
C_{5h}		E'_2		E''_1
C_{6h}		E_{2g}		E_{1g}
D_{2h}				$B_{1g}\ B_{2g}\ B_{3g}$
D_{3h}		E'	A'_2	E''
D_{4h}		$B_{1g}\ B_{2g}$	A_{2g}	E_g
D_{5h}		E'_2	A'_2	E''_1
D_{6h}		E_{2g}	A_{2g}	E_{1g}
D_{2d}		$B_1\ B_2$	A_2	E
D_{3d}			A_{2g}	E_g
D_{4d}		E_2	A_2	E_3
D_{5d}		E_{2g}	A_{2g}	E_{1g}
D_{6d}		E_2	A_2	E_5
S_4		B		E
S_6				E_g
S_8		E_2		E_3

Table 2. Continued.

		G^s	G^a	Mixed G^s/G^a
ρ_l	Off resonance	3/4	–	3/4
	On resonance	3/4	∞	$2\ (2 < \rho_l^{max} < \infty)$
Point Group				
T		E		T
T_d		$E\ \ T_2$	T_1	
T_h		E_g		T_g
O		$E\ \ T_2$	T_1	
O_h		$E_g\ \ T_{2g}$	T_{1g}	
I		H	T_1	
I_h		H_g	T_{1g}	
$C_{\infty v}$		Δ	$A_2\ (\Sigma^-)$	$E_1\ (\Pi)$
$D_{\infty h}$		Δ_g	Σ_g^-	Π_g

The same is true for the modes of T_1 symmetry in the point groups T_d, O, I and those of T_{1g} symmetry in O_h and I_h. For the modes $T_g(T_h)$ and $T(T)$, G^a and G^s are mixed and the tensor is generally asymmetric.

Mixed symmetric/antisymmetric scattering, showing anomalous polarisation, is the general rule in non-degenerate groups. This mixed scattering is generally expected to show dispersion of the value of ρ_l in the resonance region. This is because, in a non-degenerate group, a transition has only one allowed polarisation (σ say). Thus, for resonance with the 0–0 (0–1) transition, only $\boldsymbol{\alpha}_{\rho\sigma}$ ($\boldsymbol{\alpha}_{\sigma\rho}$) is non-zero ($\boldsymbol{\alpha}_{\sigma\rho}(\boldsymbol{\alpha}_{\rho\sigma}) = 0$, assuming negligible off-resonance contribution from the 0–1 (0–0) transition) and therefore $G^a = G^s$ using Eqs. (34). Equation (36) shows that the value of ρ_l is 2 in this case. However, off-resonance $G^a \rightarrow 0$ and so $\rho_l \rightarrow 3/4$. The non-symmetry aspects of anomalous and inverse polarisation are considered in Section 2.9.2.

The value of ρ_l reaches a maximum midway between the 0–0 and 0–1 energies (E_{e0} and E_{e1}) and the value of ρ_l^{max} depends on the vibrational interval and on the bandwidths of the vibronic components (*41*). Assuming equal bandwidths, $\Gamma_{e0} = \Gamma_{e1} = \Gamma$,

$$\rho_l^{max} = \frac{3}{4} + \frac{5}{4}\left[1 + \frac{(E_0 - E_1)^2}{3\,\Gamma^2}\right]\frac{|G^a|^2}{|G^s|^2} \qquad (37)$$

at $E_L = (E_1 - E_0)/2$ (see *Case B* of Section 2.9.2).

The distribution of the tensor invariants is given in Table 2 for non-totally symmetric Raman processes in the molecular point groups. The symmetry of the process is given for each point group together with the expected values of ρ_1 for the different types of scattering: – first column: pure symmetric scattering; second column: pure antisymmetric scattering; third column: mixed symmetric/antisymmetric scattering.

2.12 Harmonic Wavenumbers and Anharmonicity Constants

The observation of a large number of overtones of a fundamental under resonance Raman conditions makes it possible to determine the harmonic wavenumbers (ω_j) and the anharmonicity constants x_{jj} and x_{jk}. The vibrational term for a polyatomic species is given by the expression

$$G = \sum_j \omega_j \left(v_j + \frac{d_j}{2}\right) + \sum_j \sum_{k \geqslant j} x_{jk} \left(v_j + \frac{d_j}{2}\right)\left(v_k + \frac{d_k}{2}\right)$$

where v_j and d_j are the vibrational quantum number and degeneracy, respectively, of the j-th fundamental. The observed wavenumber, $\nu(v_1\,\nu_1)$ of any overtone of a fundamental, ν_1, is given for Stokes bands by the expression

$$\begin{aligned}\nu(v_1\,\nu_1) &= G(v_1 + m, n, p, q, \ldots j) \\ &\quad - G(m, n, p, q, \ldots j) \\ &= v_1\,\omega_1 + v_1\,x_{11}(2\,m + d_1 + v_1) \\ &\quad + v_1\left[x_{12}\left(n + \frac{d_2}{2}\right) + x_{13}\left(p + \frac{d_3}{2}\right) + \ldots x_{1j}\left(j + \frac{d_j}{2}\right)\right]\end{aligned} \qquad (38)$$

It follows that a plot of $\nu(v_1\,\nu_1)/v_1$ versus v_1 has a slope of x_{11} and a intercept of

$$\omega_1 + x_{11}(2\,m + d_1) + \left[x_{12}\left(n + \frac{d_2}{2}\right) + x_{13}\left(p + \frac{d_3}{2}\right) + \ldots x_{1j}\left(j + \frac{d_j}{2}\right)\right] \qquad (39)$$

It is commonly assumed that the terms inside the square brackets approximate to zero (as the x_{ij} terms can be positive or negative). There is no real justification for this assumption, so the ω_1 values suffer from some uncertainty in this respect. On this basis, however, ω_1 and x_{11} may be deduced from the slope, x_{11}, and intercept, $\omega_1 + x_{11}(2\,m + d_1)$, since the latter reduces to $\omega_1 + x_{11}$ with the simplification that $d_1 = 1$ if ν_1 is totally symmetric, and $m = 0$ if the fundamental (rather than a hot band) is considered; this is a good approximation in the case of a non-degenerate mode and where the data are gathered at low temperatures.

Similarly, the observed wavenumber of a combination band, $\nu(v_1\nu_1 + v_2\nu_2)$, is given by the expression:

$$\begin{aligned}\nu(v_1\nu_1 + v_2\nu_2) &= G(v_1 + m, v_2 + n, p, q, \ldots j) \\ &\quad - G(m, n, p, q, \ldots j) \\ &= v_1\omega_1 + v_2\omega_2 + v_1 x_{11}(2m + d_1 + v_1) \\ &\quad + v_2 x_{22}(2n + d_2 + v_2) \\ &\quad + x_{12}\left(v_1 n + v_2 m + v_1 v_2 + \frac{v_1 d_2}{2} + \frac{v_2 d_1}{2}\right) \qquad (40) \\ &\quad + v_1\left[x_{13}\left(p + \frac{d_3}{2}\right) + x_{14}\left(q + \frac{d_4}{2}\right) + \ldots x_{1j}\left(j + \frac{d_j}{2}\right)\right] \quad j \neq 2 \\ &\quad + v_2\left[x_{23}\left(p + \frac{d_3}{2}\right) + x_{24}\left(q + \frac{d_4}{2}\right) + \ldots x_{2j}\left(j + \frac{d_j}{2}\right)\right] \quad j \neq 1\end{aligned}$$

Hence if the observed wavenumber of $v_2\nu_2$ is subtracted from that of the corresponding combination band involving $v_2\nu_2$, the following general expression is obtained:

$$\begin{aligned}\nu(v_1\nu_1 + v_2\nu_2) - \nu(v_2\nu_2) &= v_1\omega_1 + v_1 x_{11}(2m + d_1 + v_1) \\ &\quad + v_1\left[x_{12}\left(v_2 + n + \frac{d_2}{2}\right) + x_{13}\left(p + \frac{d_3}{2}\right) + x_{14}\left(q + \frac{d_4}{2}\right)\right. \qquad (41) \\ &\quad \left. + \ldots x_{1j}\left(j + \frac{d_j}{2}\right)\right]\end{aligned}$$

A plot of $[\nu(v_1\nu_1 + v_2\nu_2) - \nu(v_2\nu_2)]/v_1$ versus v_1 thus has, with the same simplifications used for Eq. (38), the identical slope, x_{11}, as that of a plot of $\nu(v_1\nu_1)/v_1$ versus v_1, but an intercept on $v_1 = 0$ which is greater by $v_2 x_{12}$. Hence, if further progressions in ν_1 are observed, based on one or more quanta of another fundamental ν_2, it is possible to deduce a value for the anharmonicity constant x_{12}.

There are several molecules for which x_{1j} values have now been determined, the most detailed studies of this sort having been carried out on the $[Ru_2OCl_{10}]^{4-}$ and $[Re_2OCl_{10}]^{3-}$ ions, as discussed in Section 4.7.

3 Techniques

As indicated in the introduction, Ar^+ and Kr^+ ion lasers between them give a choice of about 20 different lasing lines with which to record a Raman spectrum. In many resonance Raman studies, however, and particularly those involving the plotting of excitation profiles, it is necessary to be able to record Raman spectra with a large number of lasing lines which differ from one another by only a few nanometres in wavelength. Such detailed studies have only been possible since the development of the dye laser, illustrated in Fig. 8, whose characteristic property is that its output may be tuned continuously to any wavelength that falls within the fluorescence emission curve of the dye. The most commonly used and most efficient dye is rhodamine 6 G (tuning range 575–640 nm, maximum output 4 W at 600 nm for a 21 W Ar^+ ion pump, operating all-wave). Many other dyes may be made to lase in these circumstances e.g. rhodamine B (615–665 nm), rhodamine 110 (550–590 nm) and a variety of coumarins. The drawback in the case of some of the coumarin dyes is that they are not only very short-lived (ca. 48 h in the pump beam) and very expensive, but that they require uv pumping *viz*. from Ar^{2+} 351.1 and 363.8 nm exciting lines. The attraction of the coumarin dyes is, of course their tunability in the blue region, this being unfortunately under-supplied by Ar^+ and Kr^+ laser lines. New dyes (e.g. bis-styryl and stilbene 3) are under development at the present time with the object of providing more stable, more efficient, and less expensive means for tuning the laser-beam wavelength in the blue. The emission curves of all commonly used lasing dyes are shown in Fig. 9.

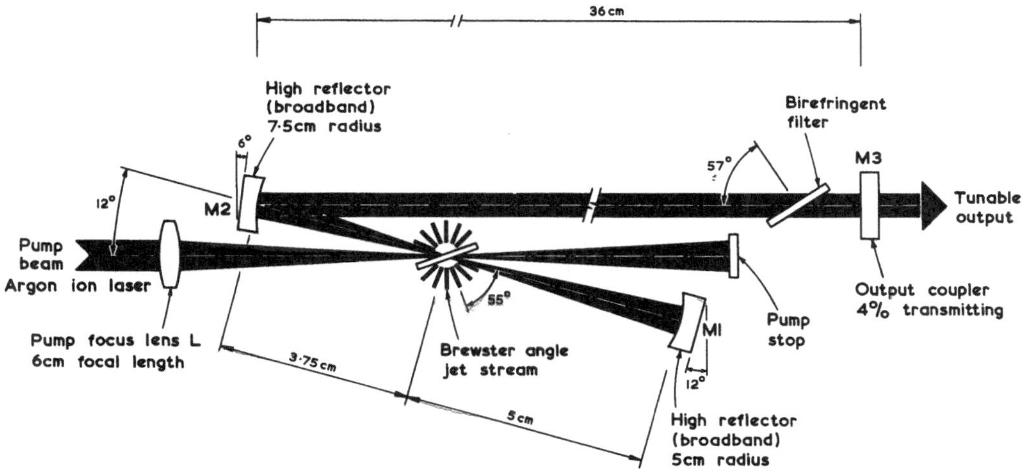

Fig. 8. Optical arrangement in a Coherent Radiation Model 490 dye laser. The pump beam is usually the all-wave emission of an argon ion laser

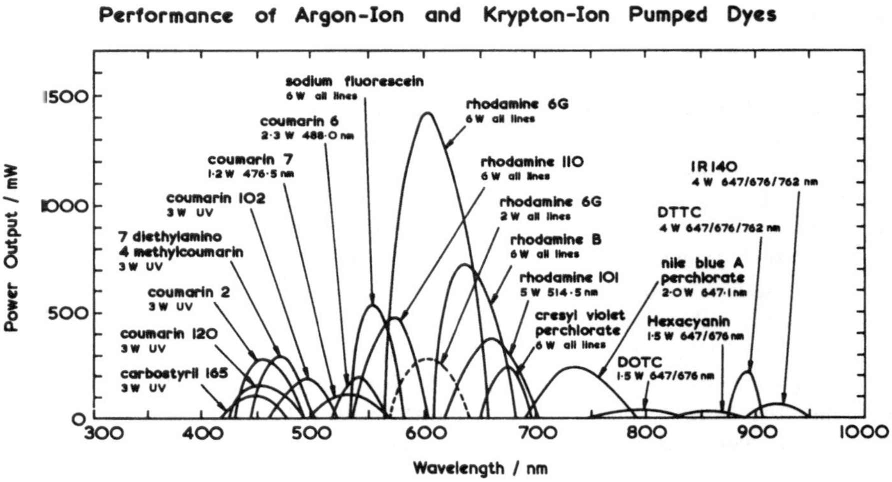

Fig. 9. Fluorescence emission curves of commonly used dyes in a dye laser. The power output (in mW) as a function of wavelength (in nm) is indicated for the specified pump line or lines (in W) for each dye

There are a number of difficulties associated with recording Raman spectra which are peculiar to resonance Raman studies. These include

(a) the optimisation of the concentration of the scattering species, i.e. the making of allowances for competition between absorption and scattering processes
(b) the avoidance of local overheating of the sample, which leads (in the case of solution studies) at best to loss of focus of the laser beam through changes in the refractive index of the medium (the thermal lens effect) and at worst to thermal decomposition of the sample and
(c) the elimination of fluorescence and photolysis.

Point (a) is best optimised by experiment, although several attempts have been made to provide a theoretical basis for determining the optimum concentration of a species of known molar decadic absorption coefficient for resonance Raman studies. The deleterious effects on the signal-noise ratio of the Raman spectrum attributable to point (b) can be obviated by spinning the sample at *ca.* 1600 rev min^{-1} as a solid, liquid or gas. Several articles contain details of many of these devices, the common object of which is to ensure relative motion between the sample and the focused laser beam (*64*). Surface scanning devices are also in use, and these enable the laser beam, while remaining focused on the surface of the sample, to scan over its surface in either a circular or linear fashion to achieve the same object. Such a procedure is particularly effective when the sample is held at liquid nitrogen temperatures. A device which allows this mode of operation is shown in Fig. 10; in this case the laser beam is refracted rapidly in a linear path over the surface of the sample, the latter being fixed to the bottom of a liquid-nitrogen *Dewar* (*65*).

The deleterious effects of photolysis on the Raman signal also seem to be largely overcome by sample spinning techniques, but those of fluorescence are only satisfactorily overcome by the use of time-resolution techniques, involving time-adjusted gate electronics to permit the separation of the resonance Raman from the fluorescence spectrum. The reader is referred elsewhere for details of the appropriate equipment (*64*).

Further important technical advances include the development of devices for the automatic scanning of the depolarisation ratio, for measuring Raman CID (circular intensity differentials), for measuring difference Raman spectra (i.e. the difference between the Raman signals from solutions and solvents), for studying optical-fibre Raman spectroscopy, for rapid (i.e. picosecond) Raman spectroscopy and Raman micrography (*64*).

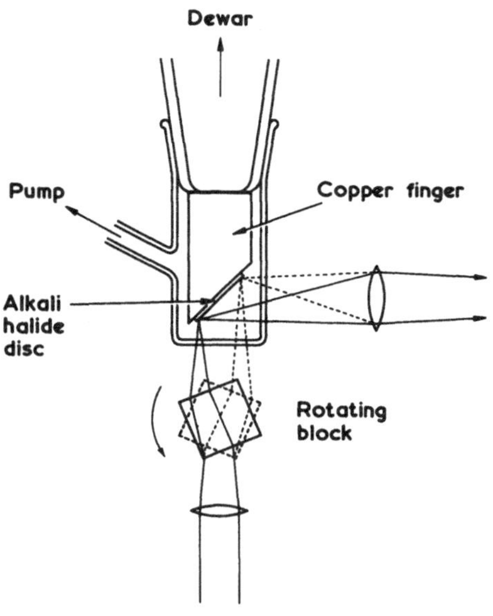

Fig. 10. Device which permits the laser beam to be flicked rapidly across the surface of a sample which is being held at ca. 80 K. [From *Clark* and *Turtle*, Ref. (*65*)]

4 Review of Experimental Results on Inorganic Systems

Earlier reviews of the results of resonance Raman studies on inorganic molecules and ions provide all the essential background information on this subject. The present review provides up-to-date tables of spectroscopic data relating to molecules studied so far and, in particular, it draws attention to the state of the species studied, the assignments of the resonant electronic transitions, the relation between the wavelengths of the resonant electronic transitions and those of the laser lines used in the studies, the harmonic wavenumbers (within the restrictions implied in Section 2.12) and anharmonicity constants derived from the wavenumbers of members of progressions in a totally symmetric mode of the scattering molecule, the lengths of the progressions, and the observed maximum values in each case of the intensity of the first overtone relative to that of the fundamental. The tables are divided according to the structural types of the species studied. Individual cases will only be discussed where they illustrate points of interest, either with respect to the theory of the subject, or with respect to particularly important areas of Inorganic Chemistry.

4.1 Diatomic Species

Diatomic cations, neutral molecules, and anions represent the type of inorganic species which has been most extensively studied by resonance Raman spectroscopy. Iodine in the gaseous, dissolved, and matrix-isolated states has been the subject of particularly detailed studies, and it is this molecule for which the greatest number of members (25) of a resonance Raman progression has so far been observed (*66*). The relation between resonance Raman and resonance fluorescence spectra has been discussed in Section 2, but it is worth illustrating the general principles involved by reference to the work on iodine.

4.1.1 Molecular Iodine

This is a good example, because (a) the electronic structure of iodine is well known from fluorescence, predissociation and photofragmentation studies and (b) iodine is mono-isotopic and homonuclear. The characteristics of the Raman spectrum of gaseous iodine depend critically on whether, with respect to the $\tilde{B}\,{}^3\Pi^+_{0u}$ state, one is dealing with normal Raman scattering, discrete resonance Raman scattering or continuum resonance Raman scattering (Fig. 3). Following *Rousseau* and *Williams* (*16*), these characteristics are as follows:

A. Normal Raman Scattering

(i) The Rayleigh line is strong and the fundamental vibrational transition is weak; overtones are usually too weak to be observed.

(ii) The band envelope of the fundamental is relatively broad, may have some structure, and displays no dependence on the excitation frequency. The structure arises from all the allowed Raman transitions from all the populated vibration-rotation levels.

(iii) The scattering intensity varies slowly and uniformly with excitation frequency (as $\nu_L{}^4$).

(iv) The Stokes/anti-Stokes ratio may be calculated from the Boltzmann factor.

(v) The scattering time is very short.

(vi) The scattering intensity is not quenched by the addition of a foreign gas.

(vii) The depolarisation ratio of the Q branch is small, the exact value being determined by the anisotropy in the polarisability.

B. Discrete resonance Raman Scattering

(i) The overtones may be of comparable intensity to the fundamental and even the Rayleigh line. The ratio of the intensity of one overtone to that of the next varies widely.

(ii) By use of sufficiently narrow excitation each overtone is seen to consist of only a few sharp lines. Their widths are determined by a combination of natural, collisional, and Doppler broadening, in addition to possible nuclear hyperfine splitting. The structure within each overtone band changes dramatically with very small changes in the excitation frequency.

(iii) The scattered intensity is strongly dependent on the excitation frequency.

(iv) The Stokes/anti-Stokes ratio is not simply related to the ground-state population factors.

(v) The scattering time is long (ca. 10^{-5}–10^{-8} s); however, if the excitation frequency is moved away from exact resonance, the scattering time becomes very short.

(vi) The addition of a foreign gas strongly quenches the intensity of the scattering.

(vii) The Q-branch scattering is generally depolarised, with a depolarisation ratio of 3/4.

C. Continuum Resonance Raman Scattering

(i) Numerous overtones occur, with intensities comparable with that of the fundamental. However, contrary to case B, the intensity variation from one overtone to the next is very regular.

(ii) The band envelope for each overtone is broad, with structure which smoothly changes with change in the excitation frequency.

(iii) The scattered intensity varies smoothly with the excitation frequency.

(iv) The Stokes/anti-Stokes ratio varies systematically with excitation frequency.

(v) The scattering time is short.

(vi) The addition of a foreign gas does not quench the scattered intensity, and
(vii) The depolarisation ratio is small and may be calculated from the electronic structure of the molecule.

For continuum resonance Raman scattering the scattered intensity is governed by the weighted sums of the Franck-Condon overlap amplitudes. For discrete resonance Raman scattering, the continuum contribution is swamped by strong resonances with one or more specific states. The intensities of the various overtones vary erratically owing to wide variations in the overlap integrals. The strong dependence of discrete resonance Raman scattering on the excitation frequency is understandable in that, by changing the frequency, resonance can thereby occur with an entirely different intermediate state. For continuum resonance Raman scattering, the spectrum also changes with excitation frequency but, owing to the averaging effects referred to above, the changes are smooth and continuous.

The very different spectra of iodine obtained under continuum and discrete resonance-Raman conditions are illustrated in Fig. 11 for resonance with the $\tilde{B}\ ^3\Pi^+_{0u}$ state, whose dissociation limit is 20,162 cm^{-1}. In the case illustrated of discrete resonance-Raman scattering, $\lambda_L = 514.5$ nm, and specific re-emission results from an initial transition from the $v'' = 1$ vibrational, $J'' = 99$ rotational level of the $\tilde{X}$ state to the $v' = 58$, $J' = 100$ level of the $\tilde{B}$ state, i.e. the transition is $58' - 1''$ R(99). Owing to the rotational selection rule for dipole radiation, $\Delta J = \pm 1$, a pattern of doublets appears in the emission. Clearly, the continuum resonance-Raman spectrum of iodine ($\lambda_L = 488.0$ nm) is very different from the discrete case spectrum. The structure, which arises from the O, Q, and S branches of the multitude of vibration-rotation transitions occurring, can be analysed in terms of a Fortrat diagram, as done for gaseous bromine (*67*).

It should be noted that, in the case of iodine as well as in that of bromine, there is strong evidence of resonance coupling of the exciting radiation with the vibronic levels of not only the $^3\Pi^+_{0u} \leftarrow {}^1\Sigma^+_g$ transition but also the $^1\Pi_{1u} \leftarrow {}^1\Sigma^+_g$ transition; approximately 30% of the total absorption intensity has been attributed to the singlet-singlet transition in iodine (*66*). Data on other diatomic species are given in Table 3.

4.1.2 Ionic Species

Continuum resonance-Raman scattering can be observed under discrete resonance-Raman scattering conditions only if the resonance fluorescence is quenched, either with an inert gas, or (in the case of condensed phase studies) by the solvent or matrix. Thus, on excitation of a liquid, solution, or solid within the contour of an absorption band, the Raman spectrum observed has the characteristics of the continuum rather than the discrete case or, in other terminology, of resonance Raman, rather than resonance fluorescence spectra. Such spectra provide unique information on the spectroscopic properties of radical cations and ions, some of which species are unstable in air. Particularly noteworthy have been the studies by *Andrews* et al. (*68*) which have

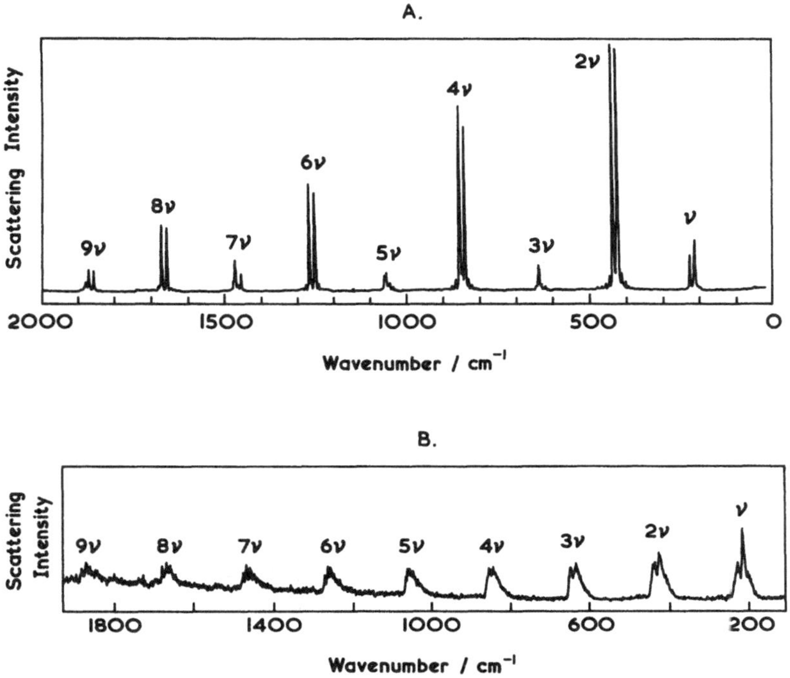

Fig. 11. A. Spectrum showing discrete resonance Raman scattering from I_2 gas. Excitation was with a single mode of the 514.5 nm Ar^+ laser line in resonance with the transition $\tilde{B}\,(^3\Pi^+_{0u})$ v' = 58, J' = 100 $\leftarrow$ $\tilde{X}\,(^1\Sigma^+_g)$ v'' = 1, J'' = 99 [Ref. (*16*)].
The doublet structure of each vibrational overtone results from the $\Delta J = \pm 1$ rotational selection rule for an electric-dipole transition.
B. Continuum resonance-Raman scattering from I_2 gas with excitation at 488.0 nm which is above the dissociation limit of the $\tilde{B}(^3\Pi^+_{0u}) \leftarrow \tilde{X}(^1\Sigma^+_g)$ transition. The fine structure of each vibrational overtone is attributable to the Q, O and S branches of the multitude of rotational transitions occurring

led to the characterisation of the radical anions Cl_2^-, Br_2^- and I_2^- in argon matrices, and, of the S_2^-, SSe^- and Se_2^- radical anions in alkali halide lattices (*69, 70*) and in ultramarine blue, green and red (*71*). Very recently, the homonuclear noble-gas cation Xe_2^+, has been characterised for the first time in a condensed phase (*72*). Its colour (green) and its spectroscopic properties are closely similar, as expected, to those of the isoelectronic species I_2^-.

Both Xe_2^+ and I_2^- have a vibrational frequency roughly half that of molecular iodine, consistent with the (molecular-orbital based) idea of a half bond for the ionic species versus a full bond for molecular iodine. Many other ionic species may be expected to be trapped in matrices and studied in the future by resonance Raman spectroscopy.

Table 3. Resonance Raman spectral data on diatomic species

Species	State	Assignment	λ_{max}/nm	$\lambda_{\mathscr{L}}$/nm	ω_e/cm^{-1}	$\omega_e x_e$/cm^{-1}	Progression	$I(2\nu_1)/I(\nu_1)$	Ref.
Br_2^+	SO_3F^- salt in $HSO_3F/SbF/$-SO_3 soln.	$\leftarrow {}^2\Pi_{3/2g}$	510	514.5	361.5 ± 1	0.9 ± 0.2	6 ν	1.0	a,b
I_2^+	HSO_3F soln.	$\leftarrow {}^2\Pi_{3/2g}$	640	632.8	238	0	4 ν	0.70	c
Xe_2^+	SbF_6^- salt in SbF_5 soln.	${}^2\Pi_{1/2g} \leftarrow {}^2\Sigma_u^+$	710	676.4	123	0 ± 1	4 ν	0.5?	d
$^{35}Cl_2$	Gas, 3 atm.	${}^3\Pi_{0u}^+ \leftarrow {}^1\Sigma_g^+$	330	363.8	559.5	2.84	8 ν	0.17	e,f
$^{79}Br_2$	Gas	${}^3\Pi_{0u}^+ \leftarrow {}^1\Sigma_g^+$	483	488.0	325.26 ± 0.10	1.070 ± 0.013	10 ν	1.07	g,h,i,j
	Ar matrix, 16 K $(Br_2)_n$		420	457.9	295 ± 1	0 ± 1	4 ν	0.21	k
$Br^{35}Cl$	Gas	${}^3\Pi_0^+ \leftarrow {}^1\Sigma^+$	–	488.0	440 ± 1		2 ν	~ 0.15	g
$I^{35}Cl$	Gas	${}^3\Pi_0^+ \leftarrow {}^1\Sigma^+$	480	488.0	381 ± 1		⩾ 5 ν	0.95	g
I Br	Gas	${}^3\Pi_0^+ \leftarrow {}^1\Sigma^+$	507	488.0	265 ± 1		⩾ 7 ν	1.08	g
I_2	Gas	${}^3\Pi_{0u}^+, {}^1\Pi_{1u} \leftarrow {}^1\Sigma_g^+$	505	488.0	214.534 ± 0.040	0.6070 ± 0.0085	16 ν	0.71	g,l
	n-Heptane soln.		521	514.5	212.51 ± 0.10	0.56 ± 0.04	20 ν	–	m
	$CHCl_3$ soln.		513	514.5	212.18 ± 0.08	0.64 ± 0.03	13 ν	0.70	m,n,o
	Ar matrix, 16K		523	530.9	213.70 ± 0.22	0.60 ± 0.01	25 ν	0.865	k,p
	Ar matrix, 16K $(I_2)_n$		523	514.5	181.0 ± 0.3	0.08 ± 0.03	11 ν	0.35	k
Sn_2	Ar matrix, 12K			488.0	188.0 ± 0.5	0.53 ± 0.05	8 ν	0.67	q
S_2^-	Doped in NaI, 80K	${}^2\Pi_u \leftarrow {}^2\Pi_g$	431	454.5	599.8	2.6	10 ν	0.83	r,s
	Ultramarine		400	457.9	594.3	2.5	5 ν		t,u

SSe^-	Doped in KI			488.0	464	± 1	2.0	± 0.3	3 ν	–	r
Se_2^-	Doped in KI			488.0	325	± 1	0.75	± 0.3	3 ν	–	r
$^{35}Cl_2^-$	Li^+ salt, Ar matrix	$^2\Sigma_g^+ \leftarrow {}^2\Sigma_u^+$	450	488.0	249.1	± 0.3	1.61	± 0.08	9 ν	~ 0.5	v
Br_2^-	Li^+ salt, Ar matrix 12K	$^2\Pi_{1/2g} \leftarrow {}^2\Sigma_u^+$	660	647.1	149	± 3	–		2 ν?	?	w
I_2^-	K^+ salt, Ar matrix 12K	$^2\Pi_{1/2g} \leftarrow {}^2\Sigma_u^+$	800	647.1	114.2		0.50		6 ν	0.28	x,y

a Gillespie, R. J., Morton, M. J.: Chem. Commun. 1565 (1968).
b Booth, M., Gillespie, R. J., Morton, M. J.: Adv. Raman Spectrosc. *1*, 364 (1973).
c Gillespie, R. J., Morton, M. J.: J. Mol. Spectrosc. *30*, 178 (1969).
d Stein, L., Norris, J. R., Downs, A. J., Minihan, A. R.: in press.
e Chang, H., Hwang, D.-M.: J. Chem. Phys. *67*, 4777 (1977).
f Ault, B. S., Howard, W. F., Andrews, L.: J. Mol. Spectrosc. *55*, 217, (1975).
g Holzer, W., Murphy, W. F., Bernstein, H. J.: J. Chem. Phys. *52*, 399, (1970).
h Berjot, M., Jacon, M., Bernard, L.: Compt. Rend. *B274*, 1274 (1972).
i Kiefer, W., Schrötter, H. W.: J. Chem. Phys. *53*, 1612, (1970).
j Baierl, P., Kiefer, W.: J. Raman Spectrosc. *3*, 353 (1975); $\omega_e y_e = -0.0025 \pm 0.0015\ cm^{-1}$.
k Howard, W. F., Andrews, L.: J. Raman Spectrosc. *2*, 447, (1974).
l Kiefer, W., Bernstein, H. J.: J. Mol. Spectrosc. *43*, 366, (1972).
m Kiefer, W., Bernstein, H. J.: J. Raman Spectrosc. *1*, 417, (1973).
n Mortensen, O. S.: J. Mol. Spectrosc. *39*, 48 (1971).
o Kiefer, W., Bernstein, H. J.: Appl. Spectrosc. *25*, 500, (1971).
p Grzybowski, J. M., Andrews, L.: J. Raman Spectrosc. *4*, 99, (1975); $\omega_e y_e = -0.019 \pm 0.002\ cm^{-1}$.
q Teichmann, R. A., Epting, M., Nixon, E. R.: J. Chem. Phys. *68*, 336, (1978).
r Holzer, W., Murphy, W. F., Bernstein, H. J.: J. Mol. Spectrosc. *32*, 13, (1969).
s Sawicki, C. A., Fitchen, D. B.: Chem. Phys. Letters *40*, 420 (1976); J. Chem. Phys. *65*, 4497, (1976).
t Clark, R. J. H., Franks, M. L.: Chem. Phys. Letters *34*, 69, (1975).
u Clark, R. J. H., Cobbold, D. G.: Inorg. Chem., *17*, 3169 (1978).
v Howard, W. F., Andrews, L.: J. Amer. Chem. Soc. *95*, 2056 (1973); Inorg. Chem., *14*, 767, (1975). $Li^+Cl_2^-$ may be better regarded as a trinuclear ion pair with a triangular geometry.
w Wight, C. A., Ault, B. S., Andrews, L.: in press.
x Howard, W. F., Andrews, L.: J. Amer. Chem. Soc., *97*, 2956, (1975).
y Andrews, L.: J. Amer. Chem. Soc., *98*, 2152, (1976).

Table 4. Resonance Raman spectral data on triatomic species

Species	State	Assignment	λ_{max}/nm	$\lambda_{\mathscr{L}}$/nm	ω_1/cm^{-1}	x_{11}/cm^{-1}	Progression	$I(2\nu_1)/I(\nu_1)$	Ref.	Other Progns
NO_2	Gas, 3 atm SF_6	$\tilde{B}\,^2B_1 \leftarrow \tilde{X}\,^2A_1$	~ 435	488.0	1325.64 ± 0.60	− 5.836 ± 0.033	$5\nu_1$	0.78	a,b,c,d	$2\nu_2, \nu_1 + 6\nu_2, \nu_2 + 4\nu_1$ etc.
	Gas, 3 atm SF_6	$\tilde{A}\,^2B_2 \leftarrow \tilde{X}\,^2A_1$	~ 500	496.5	750.18 ± 0.03	− 0.497 ± 0.005	$5\nu_2$	~ 0.4	a,b,c,d	
	Ar matrix, 16K		Orange-brown	488.0	1331 ± 2	− 3 ± 2	$2\nu_1$	0.25	e	$3\nu_2, \nu_1 + \nu_2$
ClO_2	$CFCl_3$ soln.	$\tilde{A}\,^2A_2 \leftarrow \tilde{X}\,^2B_1$	362	457.9	951.2	− 4.8	$4\nu_1$	0.51	f	$\nu_2 + 2\nu_1$
	Ar matrix, 16K		~ 360	457.9	949.1 ± 0.3	− 3.55 ± 0.05	$6\nu_1$	~ 0.2	g	
$^{16}O_3^-$	Na^+ salt, Ar matrix, 16K	$\tilde{A}\,^2A_2 \leftarrow \tilde{X}\,^2B_1$	~ 440	488.0	1021.3 ± 1	− 5.33 ± 0.5	$3\nu_1$	~ 1.4	h	
	Cs^+ salt, Ar matrix, 16K		~ 440	488.0	1028.2 ± 1.0	− 4.95 ± 0.25	$4\nu_1$	0.67	h	
$^{18}O_3^-$	Cs^+ salt, Ar matrix, 16K	$\tilde{A}\,^2A_2 \leftarrow \tilde{X}\,^2B_1$	~ 440	488.0	970.7 ± 0.3	− 4.42 ± 0.10	$5\nu_1$	0.50	h	
S_3^-	Doped in NaCl	$^2A_2 \leftarrow {}^2B_1$	610	488.0	532	− 1	$6\nu_1$	0.46	i,j,k	
	Ultramarine blue		610	647.1	550.3 ± 0.3	− 0.75 ± 0.25	$6\nu_1$	0.51	l	$\nu_2 + 3\nu_1$
	Dimethylformamide soln.		600	647.1	537.5	− 1.3	$6\nu_1$	0.69	m	$\nu_2 + 3\nu_1$
Br_3^-	$(SNBr_{0.4})_x$ Solid, 80K	$\sigma^* \leftarrow \sigma_g$?	260?	457.9	157	~ 0.0	$6\nu_1$	0.28	n	
I_3^-	K^+ salt in CH_3OH soln.	$\sigma^* \leftarrow \sigma_g$	360	337.1	112	–	$7\nu_1$	~ 0.6	o	
	K^+ salt in CH_3OH soln.		350	363.8	111 ± 2	+ 0.6 ± 0.3	$8\nu_1$	0.59	p	
	CsI, 8K			457.9	111	~ 0	$17\nu_1$	≥ 1	q	

[a] Marsden, M. J., Bird, G. R.: J. Chem. Phys. *59*, 2766 (1973).
[b] Bird, G. R., Marsden, M. J.: J. Mol. Spectrosc. *50*, 403 (1974).
[c] Bist, H. D. Brand, J. C. D.: J. Mol. Spectrosc. *62*, 60 (1976).
[d] Bist, H. D., Brand, J. C. D., Vasudev, R,: J. Mol. Spectrosc. *66*, 399 (1977).
[e] Tevault, D. E. Andrews, L.: Spectrochim. Acta *30A*, 969 (1974).
[f] Eysel, H. H., Bernstein, H. J.: J. Raman Spectrosc. *6*, 140 (1977).
[g] Chi, F. K., Andrews, L.: J. Mol. Spectrosc. *52*, 82 (1974).
[h] Andrews, L., Spiker, R. C.: J. Chem. Phys. *59*, 1863 (1973).
[i] Holzer, W., Murphy, W. F., Bernstein, H. J.: J. Mol. Spectrosc. *32*, 13 (1969).
[j] Holzer, W., Murphy, W. F., Bernstein, H. J.: Chem. Phys. Letters *4*, 641 (1970).
[k] Holzer, W., Racine, S., Cipriani, J.: Adv. Raman Spectrosc. *1*, 393 (1973).
[l] Clark, R. J. H., Franks, M. L.: Chem. Phys. Lett. *34*, 69 (1975).
[m] Clark, R. J. H., Cobbold, D. G.: Inorg. Chem., *17*, 3169 (1978).
[n] Temkin, H., Street, G. B.: Solid State Comm. *25*, 455 (1978).
[o] Kaya, K., Mikami, N., Udagawa, Y., Ito, M.: Chem. Phys. Lett., *16*, 151 (1972).
[p] Kiefer, W., Bernstein, H. J.: Chem. Phys. Lett., *16*, 5 (1972).
[q] Martin, T. P.: Phys. Rev. B., *13*, 3617 (1976).

4.2 Triatomic Species

A summary of the resonance Raman spectral data obtained to date on triatomic species is given in Table 4. The case of NO_2 is an immensely complicated one owing to overlap of the $\tilde{A}\ ^2B_2$ and $\tilde{B}\ ^2B_1$ states, and coupling of each with the $\tilde{C}\ ^1A_2$ and ground states (*73*). At 20,000 cm^{-1} the average density of vibrational levels is about $0.4/cm^{-1}$ corresponding to a rovibrational density in excess of $10^2/cm^{-1}$.

The isolation and study of triatomic radical anions is also a matter of considerable interest. The ozonide ion has been successfully studied as alkali metal salts in an argon matrix (*74*) and the S_3^- ion as a substitutional species in a sodium chloride crystal as host (*69, 75*). The latter species has also been identified by resonance Raman studies as the one responsible for the blue colour of ultramarine blue and lapis lazuli (*70*). Sulphur also forms deep blue solutions under certain circumstances in many other media (e.g. hexamethylphosphoramide and dimethylformamide), and it is unquestionably the S_3^- ion which is the species responsible for the colour in these cases also. Note that the ion itself cannot be isolated as a salt owing to its rapid dimerisation to the S_6^{2-} ion. The S_3^- ion has also been recognised by resonance Raman studies to be present in ultramarine green and red (*71*). Clearly there is great potential in the technique with respect to the detection and characterisation of other strongly coloured radical anions e.g. SO_2^-.

4.3 Tetrahedral Species

Resonance Raman spectral results on tetrahedral species are given in Table 5. The oxyanions and the analogous sulphur species are very polarisable, and therefore give rise to strong Raman spectra even off resonance. Most of the species of this sort which have been studied (e.g. MnO_4^- and CrO_4^{2-}) have an intense, $d_\pi (M) \leftarrow p_\pi (L)$ type, electronic transition in the visible or near uv regions, and irradiation within the contour of this band leads to intense resonance Raman spectra. Long progressions in $\nu_1 (A_1)$ are observed in each case, together with much weaker, subsidiary progressions of the sort $\nu_n + v_1 \nu_1$. The resonance Raman spectrum of $[n\text{-}C_4H_9]MnO_4$ at ca. 80 K is especially intricate, ten progressions having been discerned, in all of which it is $\nu_1 (A_1)$ which acts as the progression-forming mode (Table 6) (*76*).

Neutral tetrahalides also give rise to intense resonance Raman spectra, that of titanium tetraiodide being particularly simple and extended (Fig. 12). Other tetrahalides are expected likewise to yield good resonance Raman spectra, especially when further uv excitation lines become available.

The depolarisation ratio of the $\nu_1 (A_1)$ band of the $FeBr_4^-$ ion at resonance is 0.15 rather than zero. Although the reason for this is not entirely certain, it is undoubtedly relevant that the ground state is not totally symmetric ($^6A_1$) (*29*).

Table 5. Resonance Raman spectral data on tetrahedral species

Species	State	Assignment*	λ_{max}/nm	$\lambda_{\mathscr{L}}$/nm	ω_1/cm^{-1}	x_{11}/cm^{-1}	Progression	$I(2\nu_1)/I(\nu_1)$	Ref.
MnO_4^-	K^+ salt	$^1T_2 \leftarrow {}^1A_1$	528	514.5	845.5 ± 0.5	– 1.1 ± 0.2	$8\nu_1$	~ 0.63	a
	Bu_4N^+ salt, 80K		528	476.5	836 ± 1	– 1.0 ± 0.5	$10\nu_1$	~ 0.7	b
	H_2O soln		528	514.5	839.5 ± 0.5	– 1.0 ± 0.2	$8\nu_1$	~ 0.44	a
MnO_4^{2-}	Doped in CsI, 8K	$^2T_1, {}^2T_2 \leftarrow {}^2E$	~ 580	514.5	806.3 ± 0.1	– 1.7 ± 0.1	$6\nu_1$	0.32	c
CrO_4^{2-}	K^+ salt	$^1T_2 \leftarrow {}^1A_1$	372	363.8	854.4 ± 0.5	– 0.71 ± 0.1	$10\nu_1$	0.49	a,d
	K^+ salt, disc			351.1	854.3 ± 0.5	– 0.69 ± 0.1	$11\nu_1$	0.24	d
MoS_4^{2-}	NH_4^+ salt in H_2O/NaOH	$^1T_2 \leftarrow {}^1A_1$	468	476.5	~ 465	~ 0	$5\nu_1$	~ 0.45	e
	H_2O soln		468	465.8	454.0	~ 0.0	$6\nu_1$	0.50	f,g
PS_4^{3-}	Cu^+ salt		495	496.5	393.5	– 1.15 ± 0.15	$4\nu_1$	0.03	m
$MoSe_4^{2-}$	$[(CH_3)_4N]^+$ salt in H_2O/NaOH soln	$^1T_2 \leftarrow {}^1A_1$	563	568.2	265 ± 1	0 ± 0.5	$4\nu_1$	~ 0.42	g
WSe_4^{2-}	NH_4^+ salt in H_2O/KOH soln	$^1T_2 \leftarrow {}^1A_1$		488.0	279.5 ± 1	0 ± 0.5	$4\nu_1$	~ 0.35	g
$WOSe_3^{2-}$	Cs^+ salt, H_2O/KOH soln	$^1E \leftarrow {}^1A_1$	452	488.0	280 ± 2	0 ± 1	$3\nu_1$	~ 0.27	h
VCl_4	CCl_4 soln	$^2T_1, {}^2T_2 \leftarrow {}^2E$	408	457.9	385	–	$3\nu_1$	~ 0.4	i
$FeCl_4^-$	Et_4N^+ salt, KCl disc	$^6T_2 \leftarrow {}^6A_1$	364	363.8			$2\nu_1$	0.50	d
	$MeNO_2$ soln		364	363.8	335.4	– 0.6	$4\nu_1$	0.28	d
$FeBr_4^-$	Et_4N^+ salt	$^6T_2 \leftarrow {}^6A_1$	472	476.5	202.0 ± 0.2	– 0.38 ± 0.05	$7\nu_1$	0.45	j
FeI_4^-	Et_4N^+ salt, CsI disc	$^6T_2 \leftarrow {}^6A_1$	699	647.1	141.3 ± 0.3	– 0.3 ± 0.1	$3\nu_1$	0.41	j
TiI_4	Solid	$^1T_2 \leftarrow {}^1A_1$	515	514.5	161.0 ± 0.2	– 0.11 ± 0.03	$12\nu_1$	0.61	k
	C_6H_{12} soln		515.5	514.5	161.5 ± 0.2	– 0.11 ± 0.03	$13\nu_1$	0.61	k
SnI_4	C_6H_{12} soln	$^1T_2 \leftarrow {}^1A_1$	365	363.8	151.2 ± 0.5	– 0.02 ± 0.03	$15\nu_1$	~ 0.85	d,l

* The resonant electronic transition arises from a ligand-to-metal charge transfer transition in each case *viz.* $d_\pi(M) \leftarrow p_\pi(L)$.

a Kiefer, W., Bernstein, H. J.: Mol. Phys. *23*, 835 (1972).
b Homborg, H., Preetz, W.: Spectrochim. Acta *32A*, 709 (1976).
c Martin, T. P., Onari, S.: Phys. Rev. B., *15*, 1093 (1977).
The resonance Raman spectrum of this ion has also been recorded as a 1% solid solution in KI; with 608 nm excitation the ratio $I(2\nu_1)/I(\nu_1)$ is reported to be 1.0. Chao, R.S., Khanna, R.K., Lippincott, E.R.: J. Raman Spectrosc. *3*, 121 (1975).
d Clark, R. J. H., Dines, T. J.: unpublished work.
e Ranade, A., Stockburger, M.: Chem. Phys. Lett. *22*, 257 (1973).
f Clark, R. J. H., Franks, M. L.: unpublished work.
g Königer-Ahlborn, E., Müller, A.: Spectrochim. Acta *33A*, 273 (1977).
h Müller, A., Ahlborn, E.: Spectrochim. Acta *31A*, 75 (1975).
i Kamisuki, T., Maeda, S.: Chem. Phys. Lett. *21*, 330 (1973).
j Clark, R. J. H., Turtle, P. C.: J. Chem. Soc. Faraday II *72*, 1885 (1976).
Clark, R. J. H , Turtle, P. C.: unpublished work.
k Clark, R. J. H., Mitchell, P. D.: J. Amer. Chem. Soc. *95*, 8300 (1973); J. Raman Spectrosc. *2*, 399 (1974).
l Clark, R. J. H., Mitchell, P. D.: Chem. Comm., 762 (1973).
m Sala, O., Temperini, M. L. A.: Chem. Phys. Lett. *36*, 652 (1975). The quoted values for ω_1 and x_{11} are derived from a least squares analysis of the published data.

Table 6. Progressions observed in the resonance Raman spectrum of $(n{-}C_4H_9)MnO_4$ at ca. 80K

$v_1\nu_1$	$v_1 = 1, \ldots 10$
$v_1\nu_1 + \nu_2$	$v_1 = 0, \ldots 7$
$v_1\nu_1 + \nu_3$	$v_1 = 0, \ldots 6$
$v_1\nu_1 + \nu_4$	$v_1 = 0, \ldots 9$
$v_1\nu_1 + \nu_2 + \nu_4$	$v_1 = 0, \ldots 5$
$v_1\nu_1 + 2\nu_2$	$v_1 = 0, \ldots 6$
$v_1\nu_1 + 2\nu_3$	$v_1 = 0, \ldots 5$
$v_1\nu_1 + 2\nu_4$	$v_1 = 0, \ldots 6$
$v_1\nu_1 - 2\nu_2$	$v_1 = 6$ or 7
$v_1\nu_1 - 2\nu_4$	$v_1 = 5$

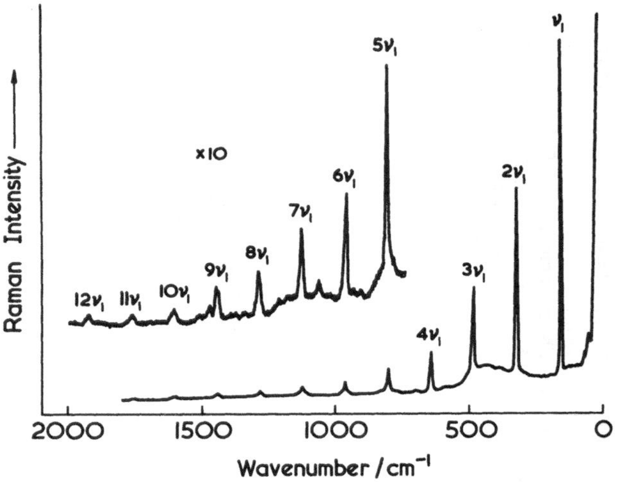

Fig. 12. Resonance Raman spectrum of solid titanium tetraiodide λ_L = 514.5 nm, power 1 W, T ~ 293 K, spectral slit width 3 cm^{-1}. [From *Clark* and *Mitchell*, J. Amer. Chem. Soc. *95*, 8300 (1973)]

4.4 Square Planar and Octahedral Species

Resonance Raman spectral results on square planar and octahedral species are given in Table 7. They are similar in most aspects to those obtained for tetrahedral species.

Certain unusual features are apparent in the resonance Raman spectra of mixed-valence species of the sort Cs_2SbCl_6, the latter being more correctly written as $Cs_4[Sb^{III}Cl_6][Sb^{V}Cl_6]$. These are the observation of

(a) a progression in $\nu_1 (A_{1g})$ of the $[SbCl_6]^-$ ion based on one quantum of a lattice-mode and

(b) a progression in $\nu_1 (A_{1g})$ of the $[SbCl_6]^-$ ion based on one quantum of $\nu_1 (A_{1g})$ of the $[SbCl_6]^{3-}$ ion.

The results have been discussed in terms of dimensional changes occurring in the lattice consequent upon excitation of the complex from the ground to the mixed-valence excited state (*77*).

The resonance Raman spectra of the $[IrCl_6]^{2-}$ and $[IrBr_6]^{2-}$ ions are important in that several bands of each are anomalously polarised. Although this discovery was initially made early in 1972 (*78*), its significance was not appreciated at the time. Independently, *Hamaguchi* and *Shimanouchi* (*61,62*) reported the results in detail and interpreted them as being a consequence of the degenerate ground state in each case; moreover, they related the specific values of the depolarisation ratio to the symmetry of the excited electronic states.

Table 7. Resonance Raman spectral data on square planar and octahedral species

Species	State	Assignment		λ_{max}/nm	$\lambda_{\mathscr{L}}$/nm	ω_1/cm^{-1}	x_{11}/cm^{-1}	Progression	$I(2\nu_1)/I(\nu_1)$	Ref.
Te_4^{2+}	Conc. H_2SO_4 soln.			510	514.5	220.5 ± 1	− 1.0 ± 0.5	$3\nu_1$	~ 0.18	a
$AuBr_4^-$	Et_4N^+ salt	$^1A_{2u}, {}^1E_u \leftarrow {}^1A_{1g}$	$d_{\sigma *}(Au) \leftarrow p_\pi(Br)$	~ 445	457.9	213.4 ± 0.5	− 0.29 ± 0.05	$9\nu_1$	0.84	b
	K^+ salt ($2H_2O$)			~ 435	457.9	210.5 ± 0.5	~ 0	$5\nu_1$	–	b
$PdBr_4^{2-}$	K^+ salt in 2M HBr soln.	$^1E_u \leftarrow {}^1A_{1g}$	$d_\pi(Pd) \leftarrow p_\pi(Br)$	332.3	325.0	–	–	$3\nu_1$	–	c
	K^+ salt in KBr disc			~ 332	325.0	190 ± 2	− 0.6 ± 0.6	$5\nu_1$	~ 0.55	c
$SbCl_6^-$	$Cs_4[Sb^{III}Cl_6][Sb^VCl_6]$		$Sb^V \leftarrow Sb^{III}$	~ 500	514.5	324 ± 1	~ 0	$4\nu_1$	0.22	d
$SbBr_6^-$	Et_4N^+ salt	$^1T_{1u} \leftarrow {}^1A_{1g}$	$a_{1g}(s_{\sigma *}) \leftarrow t_{1u}(p_\pi)$	465	457.9	191.6 ± 0.2	− 0.05 ± 0.01	$9\nu_1$	0.19	e
$IrCl_6^{2-}$	Bu_4N^+ salt	$E'_u, U'_u(^2T_{1u})$-$\leftarrow E''_g(^2T_{2g})$	$t_{2g}(d_\pi) \leftarrow t_{1u}(p_{\sigma+\pi})$	495	488.0	341.4 ± 0.5	0.0 ± 0.3	$7\nu_1$	0.79	f
$IrBr_6^{2-}$	Bu_4N^+ salt	$E''_u(^2T_{2u})$-$\leftarrow E''_g(^2T_{2g})$	$t_{2g}(d_\pi) \leftarrow t_{2u}(p_\pi)$	604	600	210.3 ± 0.5	0.0 ± 0.5	$6\nu_1$	0.69	f
$OsBr_6^{2-}$	Bu_4N^+ salt	$2\,T_{1u} \leftarrow A_{1g}$	$t_{2g}(d_\pi) \leftarrow t_{2u}(p_\pi)$	402	457.9	211.4 ± 0.5	0.0 ± 0.4	$7\nu_1$	0.92	f
$PbCl_6^{2-}$	$[Rh(NH_3)_6]_2[Pb^{II}Cl_6]$-$[Pb^{IV}Cl_6]$		$Pb^{IV} \leftarrow Pb^{II}$	~ 520	514.5	287 ± 1	− 1 ± 1	$3\nu_1$	~ 0.1	d
$PtBr_6^{2-}$	n−Bu_4N^+ salt	$^1T_{1u} \leftarrow {}^1A_{1g}$	$e_g(d_{\sigma *}) \leftarrow t_{1u}(p_\pi)$	305	325.0	209.1 ± 0.5	− 0.1 ± 0.1	$7\nu_1$	~ 0.5	g,h
PtI_6^{2-}	H_2O soln.	$^3T_{2u}, {}^3T_{1u} \leftarrow {}^1A_{1g}$	$e_g(d_{\sigma *}) \leftarrow t_{1u}(p_\pi)$	495	488.0	150.3 ± 0.5	~ 0	$3\nu_1$	~ 0.2	g,h
$[Bu_4N]$-$[OsBr_4(CO_2]$	KBr disc, 80K			510	496.5	209.5	0.15	$10\nu_3$	0.84	i
$[Bu_4N]$-$[OsI_4(CO)_2]$				510	501.7	151.7	0.3	$5\nu_3$*		i

* ν_3 is the symmetric stretching mode of the trans halogens; in the case of the bromide, additional progressions are observed which reach as far as $\nu_1 + 3\nu_3$ and $\nu_2 + 5\nu_3$, where ν_1 and ν_2 are the CO and OsC A_{1g} stretching modes respectively.

a Booth, M., Gillespie, R. J., Morton, M. J.: Adv. Raman Spectrosc. *1*, 364 (1973).
b Bosworth, Y. M., Clark, R. J. H.: Chem. Phys. Lett. *28*, 611 (1974); J. Chem. Soc. (Dalton) 381 (1975).
c Hamaguchi, H., Harada, I., Shimanouchi, T.: Chem. Lett. 1049 (1973). A detailed analysis of the electronic spectrum of this ion is given by Harrison, T. G., Patterson, H. H., Godfrey, J. J.: Inorg. Chem. *15*, 1291 (1976).
d Clark, R. J. H., Trumble, W. R.: J. Chem. Soc. (Dalton) 1145 (1976).
e Clark, R. J. H., Duarte, M. L.: J. Chem. Soc. (Dalton) 790 (1977).
f Clark, R. J. H., Turtle, P. C.: J. Chem. Soc. (Faraday II), *74*, 2063 (1978).
g Bosworth, Y. M., Clark, R. J. H.: J. Chem. Soc. (Dalton) 1749 (1974).
h Hamaguchi, H., Harada, I., Shimanouchi, T.: J. Raman Spectrosc. *2*, 517 (1974).
i Johannsen, F. H., Preetz, W.: Z. Naturforsch. *32b*, 625 (1977).

An electronic Raman spectrum of the $[IrCl_6]^{2-}$ ion has been obtained from a resonance Raman study of this ion (*79*). The observed spectrum arises from transitions between the two spin-orbit components (U'_g, E''_g) of the $^2T_{2g}$ electronic ground state. An electronic Raman spectrum has also been observed for the $OsBr_6^{2-}$ ion (*79*, *80*).

4.5 *Metal-Metal Bonded Species*

Among the simplest metal-metal bonded species are those of the sort $[M_2X_8]^{n-}$, with D_{4h} symmetry (*81*). The eight halogen ions bonded to the two metal ions are very nearly (within ca. 2%) situated at the corners of a cube, and $\widehat{MMX} = (105 \pm 1.3)^0$ for the $[Mo_2Cl_8]^{4-}$, $[Mo_2Br_8]^{4-}$, $[Tc_2Cl_8]^{3-}$, $[Re_2Cl_8]^{2-}$, and $[Re_2Br_8]^{2-}$ ions (Table 8). The molybdenum and rhenium species are isoelectronic, and each gives rise in the visible region to a strongly structured band, the structure arising primarily from a progression in ν(MM) in each case. This feature is well illustrated in the ca. 80 K electronic sprectra of the $[Mo_2Cl_8]^{4-}$ and $[Mo_2Br_8]^{4-}$ ions (Fig. 13). A simple mo-

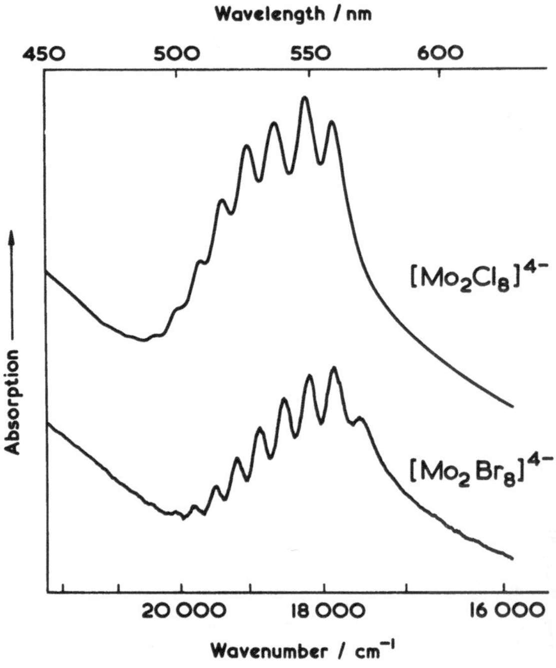

Fig. 13. Electronic spectra of the $[Mo_2Cl_8]^{4-}$ and $[Mo_2Br_8]^{4-}$ ions as KCl and KBr discs, respectively, at ca. 80 K

Table 8. Structural data on $M_2X_8^{n-}$ ions

Parameter	$[Mo_2Cl_8]^{4-}$	$[Mo_2Br_8]^{4-}$	$[Tc_2Cl_8]^{3-}$	$[Re_2Cl_8]^{2-}$	$[Re_2Br_8]^{2-}$
r(MM)/Å	2.141	2.135	2.117	2.222	2.228
$r(XX)_{ax}$/Å	3.418	3.595	3.327	3.337	3.48
$r(XX)_{equ}$/Å	3.36	3.535	3.232	3.281	3.39
$r(XX)_{ax}/r(XX)_{equ}$	1.017	1.017	1.029	1.017	1.027
r(MX)	2.457	2.604	2.364	2.320	2.478
∡ MMX/°	105.0	106.3	104.81	103.9	104.6
∡ XMX_{cis}/°	86.2	85.5	86.2	86.7	86.4
Ref.	a	b	c	d	e

[a] Brenčič, J. V., Cotton, F. A.: Inorg. Chem. *9*, 346 (1970) (average of the data for the K^+, enH_2^{2+} and NH_4^+ salts).
[b] Brenčič, J. V., Leban, I., Segedin, P.: Z. anorg. allgem. chem. *427*, 85 (1976).
[c] Cotton, F. A., Shive, L. W.: Inorg. Chem. *14*, 2032 (1975).
[d] Cotton, F. A., Frenz, B. A., Stults, B. R., Webb, T. R.: J. Amer. Chem. Soc. *98*, 2768 (1976).
[e] Cotton, F. A., De Boer, B. G., Jeremic, M.: Inorg. Chem. *9*, 2143 (1970).

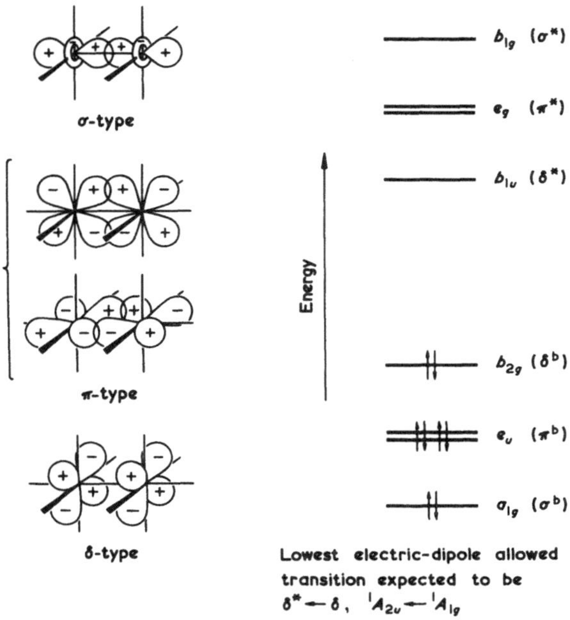

Fig. 14. Qualitative molecular orbital diagram for quadruply bonded metal-metal species for which the electronic configuration of the metal ion is d^4 (D_{4h} nomenclature)

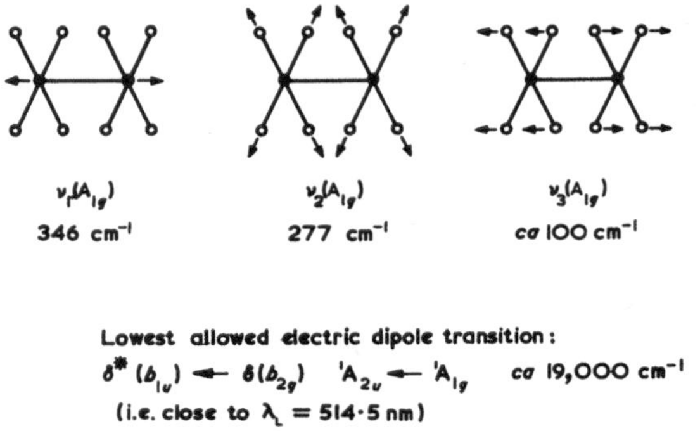

Fig. 15. Vibrational and electronic spectral data on the $[Mo_2Cl_8]^{4-}$ ion (schematic representation in which the ClMoMo is put equal to 90° rather than the correct angle of 105.0°)

lecular orbital treatment of the bonding in such ions (Fig. 14) leads to the conclusion that the metal-metal bond is quadruple ($\sigma^2 + \pi^4 + \delta^2$). The resonant electronic transition is assigned as $^1A_{2u} \leftarrow {}^1A_{1g}$, $\delta^*(b_{1u}) \leftarrow \delta(b_{2g})$, in each case (*82–84*).

The $M_2X_8^{n-}$ ions clearly possess three totally symmetric fundamentals (Fig. 15), and the key question of interest is to establish which one of these three modes, or whether more than one, displays the resonance Raman effect. In fact, irradiation within the contour of the $\delta^* \leftarrow \delta$ transition of all four ions gives rise to a long progression, and in each case it is the ν(MM) mode which acts as the progression-forming mode (Fig. 16). This result can be understood in terms of the ideas of *Tsuboi* (*85*), who has argued that the coordinates involved in converting the molecule from its

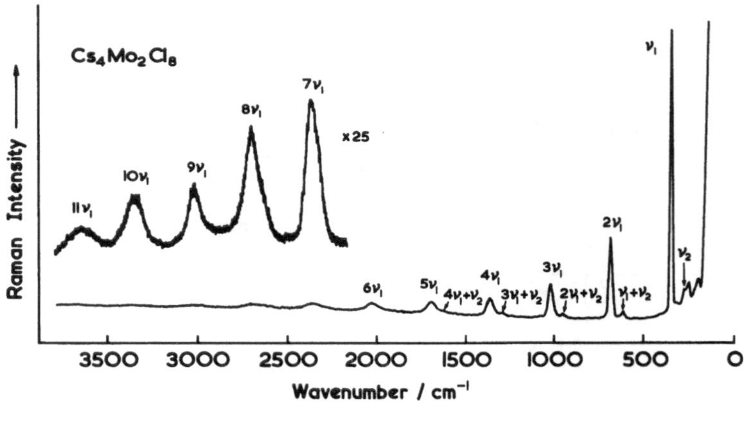

Fig. 16. Resonance Raman spectrum of $Cs_4[Mo_2Cl_8]$ at ca. 293 K. λ_L = 514.5 nm. [From *Clark* and *Franks*, Ref. (*82*)]

Table 9. Resonance Raman spectral data on metal-metal bonded species

Species	State	Assignment	λ_{max}/nm	$\lambda_{\mathscr{L}}$/nm	ω_1/cm^{-1}	x_{11}/cm^{-1}	Progressions*	$I(2\nu_1)/I(\nu_1)$	Ref.
$[Mo_2Cl_8]^{4-}$	NH_4^+ salt, NH_4Cl	${}^1A_{2u} \leftarrow {}^1A_{1g}, \delta^* \leftarrow \delta$	535	514.5	339.6 ± 0.3	− 0.76 ± 0.07	$9\nu_1, \nu_2+4\nu_1$	0.50	a,b,c
$[Mo_2Br_8]^{4-}$	NH_4^+ salt	${}^1A_{2u} \leftarrow {}^1A_{1g}, \delta^* \leftarrow \delta$	555	568.2	336.9 ± 0.4	− 0.48 ± 0.09	$6\nu_1, \nu_2+4\nu_1$	0.70	d
$[Re_2Cl_8]^{2-}$	$n-Bu_4N^+$ salt	${}^1A_{2u} \leftarrow {}^1A_{1g}, \delta^* \leftarrow \delta$	690	600	272.6 ± 0.4	− 0.35 ± 0.05	$4\nu_1, \nu_2+2\nu_1$	0.22	e
$[Re_2Br_8]^{2-}$	$n-Bu_4N^+$ salt	${}^1A_{2u} \leftarrow {}^1A_{1g}, \delta^* \leftarrow \delta$	709	647.1	276.2 ± 0.5	− 0.39 ± 0.06	$4\nu_1, \nu_2+2\nu_1$	0.31	e
$Mo_2Cl_4[P(C_4H_9)_3]_4$	Solid			514.5	352.4 ± 1.0	− 1.2 ± 0.5	$3\nu_1$	0.45	f
$Ru_2(O_2CCH_3)_4Cl$	Solid	$b_{1u} \leftarrow b_{2g}, \delta^* \leftarrow \delta$	464	514.5	327.6 ± 0.5	− 0.13 ± 0.02	$5\nu_1, \nu_2+3\nu_1$	0.15	g
$Ru_2(O_2CC_3H_7)_4Cl$	Solid	$b_{1u} \leftarrow b_{2g}, \delta^* \leftarrow \delta$	490	514.5	331.4 ± 0.5	− 0.27 ± 0.03	$4\nu_1$	0.18	g
Re_3Cl_9	Polymeric solid		515	514.5	252.4 ± 0.6	− 0.72 ± 0.08	$4\nu_1$	0.24	h
	Ar matrix (isolated)		~ 515	488.0	278.0 ± 0.5	− 0.5 ± 0.1	$5\nu_1, \nu_2+4\nu_1$ $2\nu_2+4\nu_1$ $3\nu_2+4\nu_1$	0.53	i

* ν_1 Is the metal-metal stretching mode in each case.

a Angell, C. L., Cotton, F. A., Frenz, B. A., Webb, T. R.: Chem. Comm. 399 (1973).
b Clark, R. J. H., Franks, M. L.: Chem. Comm. 316,(1974).
c Clark, R. J. H., Franks, M. L.: J. Amer. Chem. Soc. 97, 2691 (1975).
d Clark, R. J. H., D'Urso, N. R.: J. Amer. Chem. Soc. *100*, 3088 (1978).
e Clark, R. J. H., Franks, M. L.: J. Amer. Chem. Soc. *98*, 2763 (1976).
f San Fillipo, J., Sniadoch, H. J.: Inorg. Chem. *12*, 2326 (1973).
g Clark, R. J. H., Franks, M. L.: J. Chem. Soc. (Dalton) 1825 (1976).
h Clark, R. J. H., Franks, M. L.: unpublished work.
i Howard, W. F., Andrews, L.: Inorg. Chem. *14*, 1727 (1975).

ground to its (resonant) excited-state geometry are those which suffer most resonance enhancement. Clearly a $\delta^* \leftarrow \delta$ excitation of an electron would reduce the MM bond order from four to three, and thus lead to a lengthening (albeit slight, as it is merely the δ bond which is lost) of the MM bond. The MM stretching coordinate is the one that is most effective in bringing about the required structural change between the ground and excited states, and in accord with theory, this is the mode which displays the long progressions.

A second, shorter progression is observed in the resonance Raman spectrum of each ion, based on one quantum of the ν_2 (MX), A_{1g}, stretching mode in each case (see Section 2.8). However, again it is the ν_1 (MM) mode which acts as the progression-forming mode. Cross terms x_{12} have therefore been evaluated in these cases. The resonance Raman results on these ions, and on other metal-metal bonded species, are summarised in Table 9. More extensive tabulations of data on metal-metal stretching frequencies are available elsewhere (*86*). Metal-metal bond dissociation energies (D_0) for diatomic species can be estimated from the Birge-Sponer extrapolation.

$$D_0 = (\omega_e^2/4x_{11}) - \omega_e/2$$

Considerable difficulties arise when attempting to apply this expression to polyatomic molecules, owing to possible coupling between different modes of the same symmetry. Nevertheless, the data in Table 9 allow one to estimate the metal-metal bond dissociation energy in $[M_2X_8]^{n-}$ ions (*84*) to be around 500 kJ mol^{-1}, which is a very substantial value, exceeded among homonuclear units only by those of C≡C and N≡N.

4.6 Bridged Species

The bridged species which display the most remarkable resonance Raman spectra are those of the sort $[M_2OX_{10}]^{n-}$, M = Ru, Os, Re or W, X = Cl, Br or I, n = 3 or 4. The complex ions are all linear, owing to M—O—M π-bonding, with D_{4h} symmetry, and the ones which display the most spectacular spectra are the $[Ru_2OCl_{10}]^{4-}$ and $[Re_2OCl_{10}]^{3-}$ ions (*58, 87, 88*). Excitation within the contour of the lowest allowed electronic band of these ions gives rise at *ca.* 80 K to a very intense resonance Raman spectrum, (Fig. 17), consisting of eight progressions in each case in all of which it is ν_1, the ν_s(MOM) fundamental, which acts as the progression-forming mode (Table 10). The first four progressions involve totally symmetric modes only viz. ν_1, ν_s(MOM); ν_2, $\nu_s(MCl_{ax})$; ν_3, $\nu_s(MCl_{eq})$; ν_4, $\delta(OMCl_{eq})$. A further three progressions are remarkable in that they involve overtones of infrared-active fundamentals as enabling modes viz. $2\nu_9$ and $4\nu_9$, where ν_9 is the E_u-type δ(MOM) fundamental. Such enabling modes had not previously been recognised as playing important roles in resonance Raman spectroscopy (*88*). The assignments have been confirmed by the observation of the same kinds of progression in the resonance Raman spectra of the $[W_2OCl_{10}]^{4-}$, $[Re_2OCl_{10}]^{4-}$, and $[Os_2OCl_{10}]^{4-}$ ions (*88*).

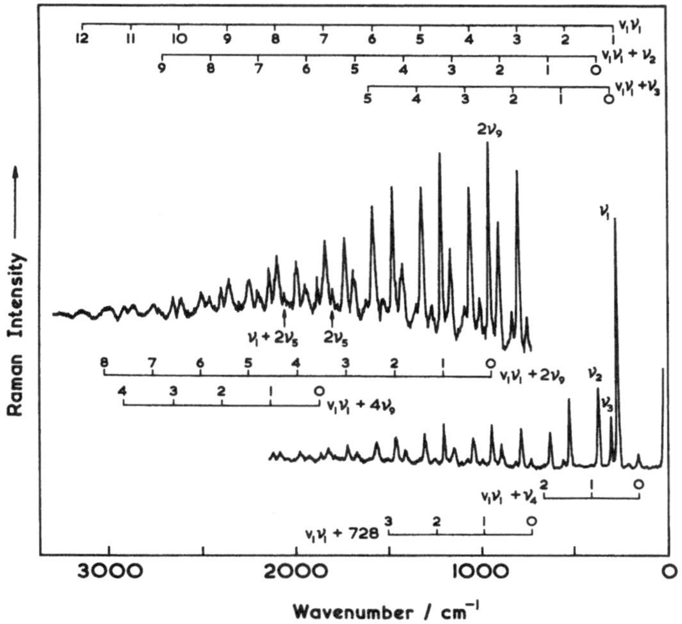

Fig. 17. Resonance Raman spectrum of $K_4[Ru_2OCl_{10}]$ at ca. 80 K, λ_L = 488.0 nm. [From *Campbell* and *Clark*, Ref. (*88*)]

Table 10. Progressions observed in the Resonance Raman Spectra of $K_4[Ru_2OCl_{10}]$ and $Cs_3[Re_2OCl_{10}]$ at ca. 80K

Progression	$K_4[Ru_2OCl_{10}]$ v_1	$Cs_3[Re_2OCl_{10}]$ v_1
$v_1\,\nu_1$	12 (ν_1 = 258.9 cm^{-1})	14 (ν_1 = 229.8 cm^{-1})
$v_1\,\nu_1 + \nu_2$	9 (ν_2 = 361.1 cm^{-1})	5 (ν_2 = 354.7 cm^{-1})
$v_1\,\nu_1 + \nu_3$	5 (ν_3 = 295.2 cm^{-1})	6 (ν_3 = 288.2 cm^{-1})
$v_1\,\nu_1 + \nu_4$	2 (ν_4 = 145.8 cm^{-1})	2 (ν_4 = 137.1 cm^{-1})
$v_1\,\nu_1 + \nu_2 + \nu_3$	– –	4 ($\nu_2 + \nu_3$ = 643.6 cm^{-1})
$v_1\,\nu_1 + 2\,\nu_5$	1 ($2\,\nu_5$ = 1779 cm^{-1})	– –
$v_1\,\nu_1 + 2\,\nu_9$	8 ($2\,\nu_9$ = 935 cm^{-1})	11 ($2\,\nu_9$ = 822.6 cm^{-1})
$v_1\,\nu_1 + 4\,\nu_9$	5 ($4\,\nu_9$ = 1858 cm^{-1})	5 ($4\,\nu_9$ = 1635 cm^{-1})
$v_1\,\nu_1$ + 728	3 (728 unassigned)	– –
$v_1\,\nu_1$ + 552.5	– –	4 (552.5 unassigned)

Studies of the excitation profiles of the totally symmetric fundamentals ν_1 to ν_4 have led to the conclusion that the resonant electronic transition is probably the $^1A_{2u} \leftarrow {}^1A_{1g}$, $e_u^* \leftarrow e_g$ transition of the Ru–O–Ru π-bond system in the case of the $[Ru_2OCl_{10}]^{4-}$ ion, and the $^4E_g \leftarrow {}^4E_u$, $e_u^* \leftarrow e_g$ transition of the Re–O–Re π-bond system in the case of the $[Re_2OCl_{10}]^{3-}$ ion.

The use of excitation profiles for the making of electronic band assignments will undoubtedly expand in the future. Promising attempts (*89*) along these lines on more extensively bridged species include those on ruthenium red, $[Ru_3O_2(NH_3)_{14}]^{6+}$, ruthenium brown $[Ru_3O_2(NH_3)_{14}]^{7+}$, and related species e.g. $[(NH_3)_5Ru–O–Ru(en)_2–O–Ru(NH_3)_5]^{6+}$, all of which have intense ($\epsilon \sim 10^5\ M^{-1}\ cm^{-1}$) electronic bands in the visible region.

4.7 Mixed-Valence Species

Mixed-valence complexes can be divided, according to the scheme of *Robin* and *Day* (*90*), into three broad categories depending on the extent of electronic interaction between the ions of different valence. In Class I complexes, the valences are firmly trapped, with no significant interaction between them, e.g. $[(Co(NH_3)_6]_2(CoCl_4)_3$, in Class II complexes, there is significant interaction, while for Class III complexes the valence delocalisation is complete, e.g. $[Nb_6Cl_{12}]^{2+}$. Thus for Class II complexes, the valences are firmly trapped with distinguishable sites, but there is sufficient overlap between orbitals on adjacent metal atoms (possibly via the intermediacy of intervening s and p orbitals of bridging halogen atoms) to permit electron transfer between the sites. Such complexes typically display intense, broad absorption bands in the visible or near infrared regions which are due to intervalence charge-transfer. Of the many mixed-valence complexes of Class II known, the halogen-bridged linear-chain complexes of platinum have thus far been the most extensively and intensively studied.

These are of the sorts (*91*):

$[Pt^{II}L_4][Pt^{IV}L_4X_2]X_4$
e.g. where L = ethylamine or propylamine, X = Cl, Br or I (Fig. 18).

$[Pt^{II}L_2X_2][Pt^{IV}L_2X_4]$
e.g. where L = ammonia, X = Cl or Br.

$[Pt^{II}(L–L)_2][Pt^{IV}(L–L)_2X_2]Y_4$
e.g. where (L–L) = 1,2-diaminoethane, or 1,2-diaminopropane, X = Cl, Br or I, Y = ClO_4

and

$[Pt^{II}(L–L)X_2][Pt^{IV}(L–L)X_4]$
e.g. (L–L) = 1,2-diaminoethane, X = Cl or Br.

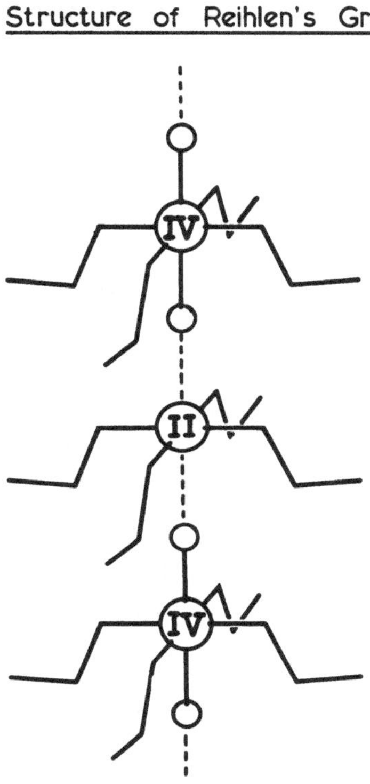

Fig. 18. Structure of Reihlen's green, $[Pt^{II}(etn)_4][Pt^{IV}(etn)_4Br_2]Br_4 \cdot 4\,H_2O$, etn = ethylamine. The analogous chloro species is known as Wolffram's red

In these complexes, the two metal-atom sites are structurally distinguishable, but may be interconverted by a concerted movement of the axial halogen atoms in phase away from the platinum (IV) atoms towards the platinum (II) atoms. Such complexes are strongly dichroic, and possess a number of highly anisotropic properties. Thus their electrical conductivity is typically about 300 times greater in the chain direction than in the perpendicular directions. The complexes are thus considered to be one-dimensional semiconductors. A hopping process has been proposed for their conductivity at ambient pressures, although at high pressures this localised process may give way to a band-type conduction process because the metal d_{z^2} and ligand p_z orbital interactions are known to increase under pressure.

Excitation of this type of mixed-valence transition within the contour of the mixed-valence band leads to a resonance Raman spectrum which is dominated by the band associated with the $\nu_1(X\text{–}Pt^{IV}\text{–}X)$ symmetric stretching fundamental, together with its associated overtone progression. This intense progression usually overwhelms the rest of the Raman-active bands, although in some cases other weak progressions have also been seen. For such subsidiary progressions, the enabling mode is another Raman-active mode, while the progression-forming mode is, as for the main progression, ν_1.

The case of Wolffram's red, $[Pt^{II}(etn)_4][Pt^{IV}(etn)_4Cl_2]Cl_4 \cdot 4H_2O$, etn = ethylamine, is typical of those for mixed-valence complexes of the sort under discussion (*92*). Excitation within the contour of the broad mixed-valence band centred around 21,000 cm^{-1} (and polarised parallel to the chain axis) leads to the appearance of a very strong band at 316 cm^{-1} (ν_1) together with its associated overtone progression (Fig. 19). This progression reaches as far as 9 ν_1 at room temperature and with 514.5 nm excitation, but as far as 16 ν_1 when the sample is held at ca. 80 K (*93*). Similar results are obtainable for the corresponding bromide (*94*).

Results on a large number of linear-chain complexes of platinum are summarised in Table 11. Harmonic wavenumbers and anharmonicity constants have been determined in all cases. The ν_1 normal coordinate seems to be related to the halogen movements involved in the proposed hopping process for the conductivity of these linear-chain mixed-valence complexes (*95*). The chain halogen atoms would need to move, on average, 0.54, 0.38 and 0.22 Å for chlorides, bromides and iodides, respectively, in order to reach the point midway between the two platinum atoms, i.e. to the situation of a platinum (III) chain. These values only differ by a factor of about two from the root-mean-square amplitudes of vibration of ν_1 in the v_1 = 16 states; these are calculated (*91*) to be 0.22 Å for X = Cl (ω_1 = 319.5 cm^{-1}) and 0.20 Å for X = Br (ω_1 = 179.6 cm^{-1}). These distance changes are related to the shift in the equilibrium

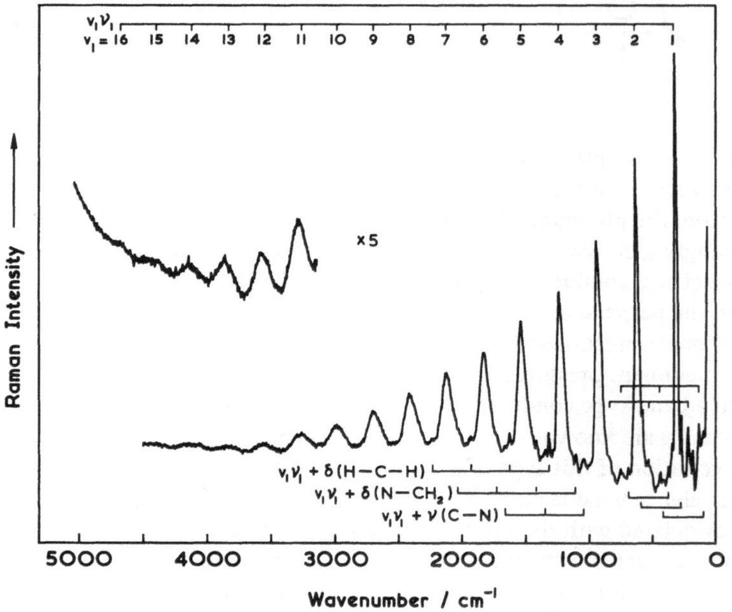

Fig. 19. Resonance Raman spectrum of Wolffram's red (see previous caption) at ca. 80 K, λ_L = 514.5 nm, spectral slit width ca. 2 cm^{-1}. [From *Clark* and *Turtle*, Ref. (*93*)]

Pt^{IV}–X bond lengths on excitation of the complex from the ground to the resonant excited state in each case. There is, however, some difficulty in calculating this shift from the theory of *Mingardi* and *Siebrand* (*96*) owing to our lack of knowledge as to the origins of the mixed-valence transitions.

Mixed-valence linear-chain complexes of palladium display very similar features to those of platinum (*97*). Indeed, the great intensity of the ν_1 progression in all these cases means that resonance Raman spectroscopy provides a sensitive and rapid means for detecting the formation of new linear-chain complexes of this sort.

Resonance Raman studies of mixed-valence complexes will undoubtedly extend much further than they have been carried to date, particularly into the realm of Class III complexes, for which a more detailed understanding of the electronic spectra is eminently desirable.

4.8 Haem Proteins and Other Biological Systems

The incorporation of metal ions into enzymes and proteins imparts specific biological activity, such as electron transfer, respiratory function, catalysis, and metal transport or storage. For transition metals, their absorption characteristics (both ligand field and charge transfer) in the visible and UV regions have permitted analysis of electronic and structural properties of use in the elucidation of metalloprotein function. Further information can be obtained from resonance Raman spectroscopy if the coupling of vibrational modes of a metal ion chromophore with either ligand field or charge transfer transitions can be understood. Little useful information of this sort has been obtained so far by excitation within the contour of ligand field bands, but excitation within the contour of charge transfer bands has proved to be most rewarding owing to the fact that bands attributable to those vibrational modes which are responsible for the coupling of the low-lying excited states of the molecule suffer resonance enhancement; thus if the vibrational band assignments are known, deductions can be drawn as to the nature of the excited electronic states.

This subject has blossomed into one in its own right, with a substantial literature with wide appeal through journals specialising in inorganic, biochemical, biophysical and theoretical chemistry. Only the barest outline may be given of work in this field owing to the confines of the space available, and the interested reader is referred in particular to a recent review by *Spiro* and *Loehr* (*98*).

Resonance Raman studies have been directed towards copper-containing proteins (e.g. haemocyanin, "blue" copper proteins), non-haem iron-containing proteins (e.g. rubredoxin, adrenodoxin, haemerythrin, transferrin) and in particular towards the haem proteins (e.g. haemoglobin, cytochrome c) and porphyrins. The electronic properties of the haems consist, in the visible and UV regions, of two π–π^* transitions of the porphyrin ring, for each of which the excited state has E_u symmetry. The higher energy transition ($\epsilon \sim 10^5$ M^{-1} cm^{-1}), near 400 nm, is called the Soret, γ or B band, whereas the lower energy one ($\epsilon \sim 10^4$ M^{-1} cm^{-1}) in the 500–600 nm re-

Table 11. Resonance Raman data on mixed-valence linear-chain complexes of platinum

Complex	Crystal colour[a])	Mixed valence band max./ cm^{-1} [b])	Excitation profile max./ cm^{-1}	Sample form	Temperature	ω_1/ cm^{-1}	x_{11}/ cm^{-1}	Progression[c])	$I(2\nu_1)/I(\nu_1)$	Ref.
Chlorides										
$[Pt(etn)_4][Pt(etn)_4Cl_2]Cl_4 \cdot 4H_2O$	Red	21,000	17,450	Powder	RT	319.5	− 2.00	$9\nu_1$	0.55	a
				KCl disc	LT	312.3	− 1.29	$16\nu_1$	1.1	b
$[Pt(en)Cl_2][Pt(en)Cl_4]$	Red	23,500	18,200	KCl disc	RT	306	–	$8\nu_1$	0.62	c
				KCl disc	LT	307.8	− 0.98	$14\nu_1$	0.96	c
$[Pt(en)_2][Pt(en)_2Cl_2](ClO_4)_4$	Red	20,000	17,000	KCl disc	LT	313.3	− 1.09	$14\nu_1$	0.61	c
$[Pt(pn)_2][Pt(pn)_2Cl_2](ClO_4)_4$	Yellow	25,000	~ 20,000	Powder	RT	315	–	$7\nu_1$	0.40	d
				NaCl disc	LT	310	− 1.0	$8\nu_1$	0.47	d
Bromides										
$[Pt(etn)_4][Pt(etn)_4Br_2]Br_4 \cdot 4H_2O$	Green	18,250	15,950	Powder	RT	179.6	− 0.37	$7\nu_1$	0.50	b,e
				KBr disc	LT	177.0	− 0.3	$12\nu_1$	~ 0.6	b
$[Pt(etn)_4][Pt(etn)_4Br_2]Br_4$	Orange	23,600	20,000	KBr disc	LT	182.5	− 0.65	$12\nu_1$	~ 0.6	b
$[Pt(en)Br_2][Pt(en)Br_4]$	Gold-green	19,000	~ 14,000	KBr disc	LT	171.7	− 0.25	$10\nu_1$	0.4	c
$[Pt(en)_2][Pt(en)_2Br_2](ClO_4)_4$	Bronze-green	18,400	17,700	$KClO_4$ disc	LT	173.2	− 0.38	$10\nu_1$	0.75	c
$[Pt(en)_2][Pt(en)_2Br_2](CuBr_2)_4$	Gold			KBr disc	RT	173	–	$6\nu_1$	0.57	d
$[Pt(pn)_2][Pt(pn)_2Br_2]Br_4$	Green	14,300	< 14,000	Powder	RT	169.8	0.0	$5\nu_1$	0.52	d
$[Pt(pn)_2][Pt(pn)_2Br_2](ClO_4)_4$	Green	18,900	< 14,000	$KClO_4$ disc	RT	176.2	− 0.54	$7\nu_1$	0.59	d
$[Pt(pn)_2][Pt(pn)_2Br_2](CuBr_2)_4$	Gold	–	–	KBr disc	RT	173	–	$6\nu_1$	0.57	d
Iodides										
$[Pt(etn)_4][Pt(etn)_4I_2]I_4$	Green-black	20,600	17,600	KI disc	LT	120.3	− 0.33	$8\nu_1$	~ 0.6	b
$[Pt(en)_2][Pt(en)_2I_2](ClO_4)_4$	Gold	14,600	14,100	$KClO_4$ disc	LT	125.3	− 1.3	$9\nu_1$	0.60	f
$[Pt(en)I_2][Pt(en)I_4]$	Bronze	15,000	⩽ 13,000	CsI disc	LT	121.2	− 0.3	$6\nu_1$	0.45	f
$[Pt(pn)_2][Pt(pn)_2I_2]ClO_4)_4$	Gold	12,500	< 12,000	Powder	RT	121.0	~ 0.0	$4\nu_1$	0.35^e	d

Abbreviations: etn = ethylamine, en = 1,2-diaminoethane, pn = 1,2-diaminopropane, RT = measured at room temperature, LT = measured in a liquid nitrogen cell.

[a]) The colours of the complexes as finely ground powders frequently differ from the crystal colour, owing to the differing relative importance of specular and diffuse reflectance.

[b]) Measured either by diffuse reflectance as a powder diluted with the appropriate alkali metal halide or by transmission as the appropriate alkali halide disc.

[c]) ν_1 Is the X–Pt^{IV}–X symmetric stretching mode.

[a] Clark, R. J. H., Franks, M. L., Trumble, W. R.: Chem. Phys. Lett. *41*, 287 (1976).

[b] Clark, R. J. H., Turtle, P. C.: Inorg. Chem., *17*, 2526 (1978).

[c] Campbell, J. R., Clark, R. J. H., Turtle, P. C.: Inorg. Chem. in press.

[d] Clark, R. J. H., Stewart, B.: to be published.

[e] Clark, R. J. H., Franks, M. L.: J. Chem. Soc. (Dalton) 198 (1977).

[f] Clark, R. J. H., Kurmoo, M., Stewart, B.: to be published.

gion is called the α or Q_0 band. This lower energy transition, however, steals about 10% of the intensity of the higher energy one through vibronic mixing, with the formation of a vibronic sideband, called the β or Q_1 band, some 1300 cm^{-1} above the α band.

Raman spectra of haems using excitation lines with $\lambda_L < 500$ nm, i.e. with lines approaching the energy of the Soret band, are dominated by bands attributable to totally symmetric modes. Thus the intense Soret band dominates the scattering mechanism, which is clearly of the A-term or FC type. For excitation with $\lambda_L > 500$ nm, the α and β bands dominate the scattering. Bands attributable to totally symmetric modes are, in this case, not detectably enhanced (see Section 2.9). Instead, bands attributable to non-totally symmetric modes are enhanced (HT scattering) owing to the strong vibronic coupling of the α to the Soret transition.

The excitation profiles of the non-totally symmetric bands reach a maximum at the maximum of the α band and a further one within the β band. The second maximum is attributed to resonance of the exciting beam with the vibronic levels that make up the β band, i.e. $\nu_L = \nu_{00} + \nu_{vib}$; accordingly the second maximum (but not the first) shifts in wavenumber depending on the vibrational wavenumber (ν_{vib}) of the Raman band in question.

The symmetries of the modes active in coupling the two E_u states are $A_{1g} + A_{2g} + B_{1g} + B_{2g}$; however, the A_{1g} modes are known to be ineffective. The B_{1g} (but apparently not the B_{2g}) modes are effective, and yield depolarised bands. The A_{2g} modes are also effective but, owing to the fact (Section 2.9.2) that their scattering tensors are antisymmetric, the resulting bands are not observed off resonance; they appear only on resonance and are then inversely polarised ($\rho_l = \infty$).

In the case of cytochrome c, no bands are inversely polarised, but some have anomalous polarisation ($3/4 < \rho_l < \infty$). This may either indicate that the A_{2g} modes are accidently degenerate with other depolarised modes, or that the effective fourfold symmetry of the haem chromophore has been reduced in cytochrome c (Fig. 20).

The α and Soret transitions are polarised in the haem plane, and hence excitation within the contours of the bands arising therefrom should alter the polarisability of the molecule in this plane leading to possible resonance enhancement to planar modes of vibration. Indeed the dominant bands in the resonance Raman spectrum arise from the in-plane porphyrin ring modes which occur in the 1100–1650 cm^{-1} region. This is consistent with the observation that the β band reaches a maximum some 1300 cm^{-1} above the α band.

Some of the porphyrin ring modes, identified via their three different states of polarisation, have been found to have wavenumbers which are sensitive to the spin-state and oxidation-state of the iron atom and thus to the structure of the porphyrin skeleton (i.e. planar or domed). Low-spin haems are six-coordinate with the iron atom in the haem plane. High-spin haems are five-coordinate or, if six-coordinate, have the sixth ligand bound less strongly than the others; in this case the iron atom lies out of the haem plane and the porphyrin ring domes, e.g. by a calculated value of 19° for deoxyhaemoglobin. The study of these spin-state and oxidation-state marker bands is of great current interest in order to obtain structural and other information on haem

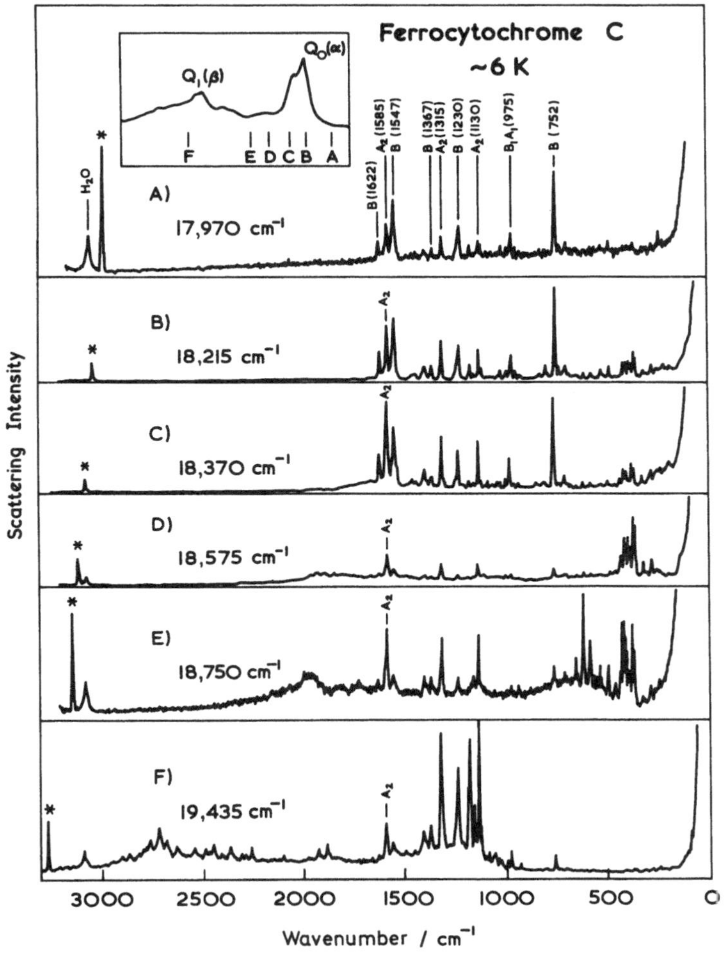

Fig. 20. Resonance Raman scattering from cytochrome c at several excitation wavenumbers. In the top spectrum the symmetries and wavenumbers of the prominent bands are labelled. For ease of identification the A_2 band at 1585 cm^{-1} is labelled in all spectra. The band labelled by an asterisk is a grating ghost. The ice line near 3100 cm^{-1} may be used as an intensity reference. The inset indicates the position of each excitation wavenumber on the absorption spectrum, which shows the $Q_0(\alpha)$ and $Q_1(\beta)$ bands. [From *Friedman* and *Rousseau*, Ref. (*110*)]

proteins e.g. on coordination by oxygen (Fig. 21). Many studies have also been carried out on other delocalised π-systems e.g. chlorophyll, vitamin B_{12}, polyolefins such as carotenoids and visual pigments, as well as nucleotide bases. The reader is referred elsewhere for information on these subjects (*98–100*).

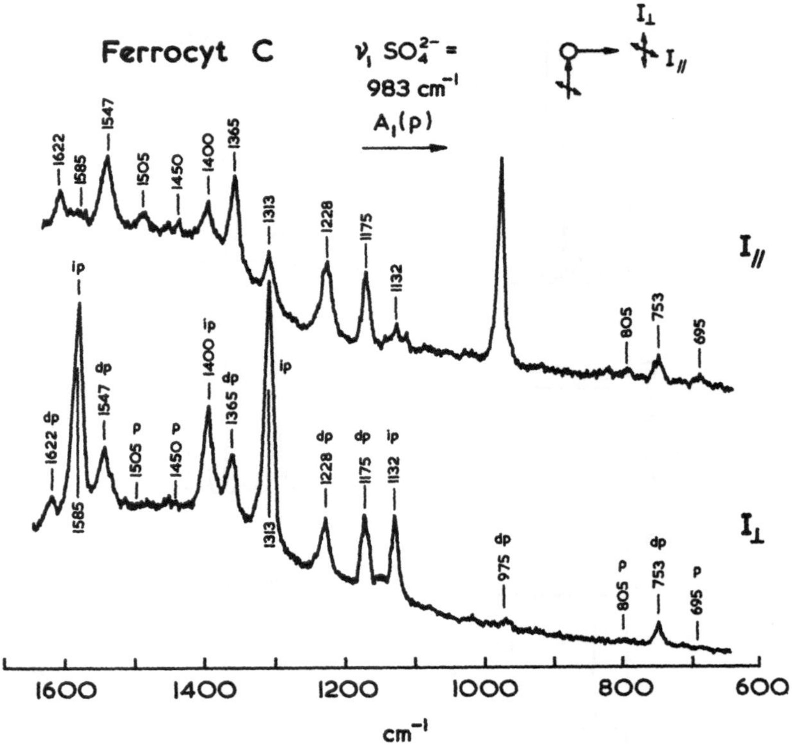
Ferrocyt C
ν_1 SO_4^{2-} =
983 cm^{-1}
A_1(p)
$I_\perp$
$I_{//}$
1622
1585
1547
1505
1450
1400
1365
1313
1228
1175
1132
805
753
695
$I_{//}$
ip
dp
p
1600
1400
1200
1000
800
600
975
$I_\perp$
cm^{-1}

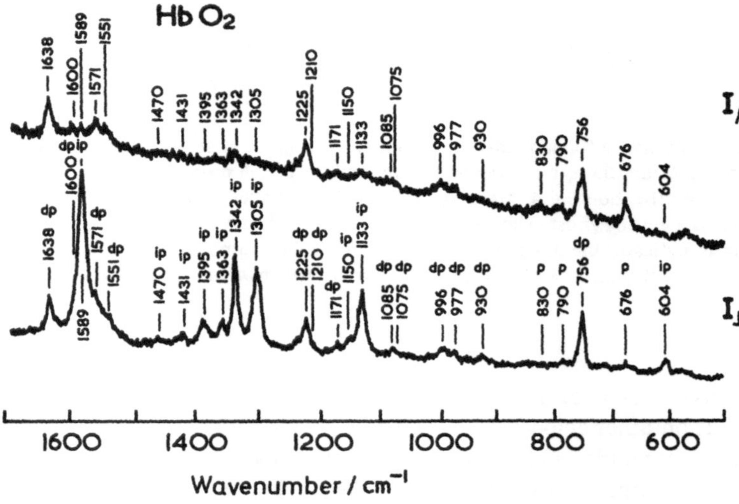
Hb O$_2$
1638
1600
1589
1571
1551
1470
1431
1395
1363
1342
1305
1225
1210
1171
1150
1133
1085
1075
996
977
930
830
790
756
676
604
$I_{//}$
$I_\perp$
ip
dp
p
1600
1400
1200
1000
800
600
Wavenumber / cm^{-1}

◁ Fig. 21. Resonance Raman spectra of oxyhaemoglobin (bottom pair of curves) and ferrocytochrome c (top pair). The scattering geometry is shown schematically in the diagram at the top. Both the direction and the polarisation vector of the incident laser radiation are perpendicular to the scattering direction. The scattered radiation is analysed into components perpendicular ($I_\perp$) and parallel ($I_\parallel$) to the incident polarisation vector. The exciting wavelength was 568.2 nm for oxyhaemoglobin and 514.5 nm for cytochrome c. The slit width was about 10 cm^{-1}. The concentrations were about 0.5 mM for each. The anomalously polarised, polarised and depolarised bands, are indicated by ip, p, and dp respectively. [From *Spiro* and *Strekas*, Ref. (*42*)]

4.9 Other Applications

The study of excitation profiles of Raman bands appears to be one of great promise. One type of application is in sorting out whether shoulders on the sides of electronic absorption bands arise from vibrational sidebands (cf. the α and β bands of haemoglobin, Section 4.8) or from additional allowed electronic bands. Such a situation arises for tris-(α-diimine) complexes of iron (II), whose spectra contain a strong band at ca. 19,000 cm^{-1} arising from an allowed charge-transfer transition from the metal ($3d$) to ligand (π) orbitals ($^1E \leftarrow {}^1A_1$), together with a shoulder some 1500 cm^{-1} to higher energy (*101*). If the unresolved shoulder were purely electronic in origin, it could only arise from $^1E \leftarrow {}^1A_2$ and/or $^1A_2 \leftarrow {}^1A_1$ transitions. Under these conditions non-totally as well as totally symmetric modes would be enhanced under resonance Raman conditions. This is because for HT coupling, terms of the type $\langle E|\partial\mathscr{H}/\partial Q|E\rangle$ are finite for A_1, A_2 and E modes and terms of the type $\langle E|\partial\mathscr{H}/\partial Q|A_2\rangle$ are finite for E modes. However, the resonance Raman spectra do not contain any inverse-polarised bands (A_2), and the only bands to show enhancement are polarised (A_1); hence it follows that the oscillator strengths of any other electronic transitions in this region are too small to account for the shoulder on the main absorption band. Thus it is attributable to a vibrational side band, and this is confirmed by the observation of substantial resonance enhancement to several ring modes (ν(C$\cdots$N)) in the 1500 cm^{-1} region of the complexes. Similar extensive studies of maleonitriledithiolato and related complexes (square planar, octahedral and trigonal prismatic) have had similar objectives (*102, 103*).

A further area of great potential is the study of short-lived free radicals. Attention has already been drawn to the stabilisation of inorganic radicals in low-temperature inert-gas matrices, in host ionic crystal lattices, or in large silicate cages (e.g. ultramarine). However, the study of electrochemically-generated transient species by *van Duyne* (*104*) and of chemically-generated radicals by *Hester* (*105*) seems to open new areas of study. Rapid mixing techniques combined with gravity-fed or pumped-flow systems are well established in the area of ESR spectroscopy, and their successful conversion to the field of Raman spectroscopy holds great promise for the detection and study of new inorganic species. The study of species adsorbed on silver, copper, platinum, gold and other electrode surfaces (*106–109*) seems to be particularly suited to resonance Raman studies, and is also a subject of considerable technological interest.

References

1. *Behringer, J.*: Raman spectroscopy, Vol. 1, p. 168. Szymanski, H. A. (ed.). New York: Plenum Press 1967
2. *Behringer, J.*: Chemical society specialist reports on molecular spectroscopy, Vol. 2, p. 100. Barrow, R. F., Long, D. A. and Millen, D. J. (eds.). London: The Chemical Society 1974
3. *Clark, R. J. H.*: Advances in infrared and Raman spectroscopy, Vol. 1, p. 143. Clark, R.J.H. and Hester, R. E. (eds.). London: Heyden 1975
4. *Kramers, H. A., Heisenberg, W.*: Z. Phys. *31*, 681 (1925)
5. *Placzek, G.*: Rayleigh and Raman Scattering. UCRL translation No. 526L from Handbuch der Radiologie Vol. 2, Marx, E. (ed.). Leipzig: Akademische Verlagsgesellschaft 1934 (a) Chap. 6, p. 50; (b) Chap. 16, p. 95; (c) Chap. 6, p. 49; (d) Chap. 6, p. 43
6. *Dirac, P. A. M.*: Proc. Roy. Soc. (London) *114*, 710 (1927)
7. *Heitler, W.*: The quantum theory of radiation. 3rd edit., p. 196. London: Oxford University Press 1954
8. *Dirac, P. A. M.*: The principles of quantum mechanics. 4th edit., p. 203. London: Oxford University Press 1958
9. *Albrecht, A. C.*: J. Chem. Phys. *34*, 1476 (1961)
10. *Albrecht, A. C., Hutley, M. C.*: J. Chem. Phys. *55*, 4438 (1971)
11. *Wilson, E. B., Decius, J. C., Cross, P. C.*: Molecular vibrations. New York: McGraw-Hill Book Company Inc. 1955
12. *Jørgensen, C. K.*: Topics Curr. Chem. *56*, 1 (1975)
13. *Herzberg, G., Teller, E.*: Z. Phys. Chem. (Leipzig) *21*, 410 (1933)
14. *Sponer, H., Teller, E.*: Rev. Mod. Phys. *13*, 75 (1937)
15. *Tang, J., Albrecht, A. C.*: Developments in the theories of vibrational Raman intensities. In: Raman spectroscopy, theory and practice, Vol. 2, pp. 33–68. Szymansky, H. A. (ed.). New York, London: Plenum 1970
16. *Rousseau, D. L., Williams, P. F.*: J. Chem. Phys. *64*, 3519 (1976)
17. *Herzberg, G.*: Spectra of diatomic molecules. Princeton N.J.: D. van Nostrand Company 1950
18. *Krushinskii, L. L., Shorygin, P. P.*: Opt. i. Spektr. *11*, 24 (1961). Optics and Spectroscopy *11*, 12 (1961)
19. *Atkins, P. W., Child, M. S., Phillips, C. S. G.*: Tables for group theory, p. 25. Oxford University Press 1970
20. *Hong, H. K.*: J. Chem. Phys. *67*, 801, 813 (1977); J. Chem. Phys. *68*, 1253 (1978)
21. *Nafie, L. A., Stein, P., Peticolas, W. L.*: Chem. Phys. Letters *12*, 131 (1971)
22. *Ballhausen, C. J., Hansen, A. C.*: Ann. Rev. Phys. Chem. *23*, 15 (1972)
23. *Rousseau, D. L., Friedman, J. M., Williams, P. F.*: The resonance Raman effect. In: Topics in Current Physics, Vol. 11, Chap. 6. Berlin, Heidelberg, New York: Springer 1978
24. *Fouche, D. G., Chang, R. K.*: Phys. Rev. Letters *29*, 256 (1972)
25. Ref. (*23*) p. 25 and *Friedman, J. M., Rousseau, D. L., Bondybey, V. E.*: Phys. Rev. Letters *37*, 1910 (1976)
26. *Martin, T. P., Onari, S.*: Phys. Rev. B*15*, 1093 (1977)
27. *Mingardi, M., Siebrand, W., Van Labeke, D., Jacon, M.*: Chem. Phys. Letters *31*, 208 (1975)
28. *Clark, R. J. H., Franks, M. L., Trumble, W. R.*: Chem. Phys. Letters *41*, 287 (1976)
29. *Clark, R. J. H., Turtle, P. C.*: J. Chem. Soc. (Faraday II) *72*, 1885 (1976)
30. *Gregory, A. R., Henneker, W. H., Siebrand, W., Zgierski, M. Z.*: J. Chem. Phys. *65*, 2071 (1976)
31. *Stein, P., Miskowski, V., Woodruff, W. H., Griffin, J. P., Werner, K. G., Gaber, B. P., Spiro, T. G.*: J. Chem. Phys. *64*, 2159 (1976)
32. *Zgierski, M. Z.*: J. Raman Spec. *6*, 53 (1977)
33. *Korenowski, G. M., Ziegler, L. D., Albrecht, A. C.*: J. Chem. Phys. *68*, 1248 (1978)
34. *Duschinsky, F.*: Acta Physiochim. USSR *1*, 551 (1973)

35. *Friedman, J.M., Hochstrasser, R.M.*: Chem. Phys. *1*, 457 (1973)
36. *Zgierski, M.Z.*: J. Raman Spec. *5*, 181 (1976)
37. *Campbell, J.R., Clark, R.J.H., Turtle, P.C.*: Inorg. Chem., in press
38. *Tsuboi et al.*: Ref. (*85*)
39. *Walsh, A.D.*: J. Chem. Soc. 2260–2331 (1953)
40. *Shelnut, J.A., Cheung, L.D., Chang, R.C.C., Nai-Teng Yu, Felton, R.H.*: J. Chem. Phys. *66*, 3387 (1977)
41. *Mortensen, O.S.*: Chem. Phys. Letters *30*, 406 (1975)
42. *Spiro, T.G., Strekas, T.C.*: Proc. Nat. Acad. Sci. U.S.A. *69*, 2622 (1972)
43. *Zgierski, M.Z.*: Chem. Phys. Letters *36*, 390 (1975)
44. *Hassing, S., Mortensen, O.S.*: Chem. Phys. Letters *47*, 115 (1977)
45. *Friedman, J.M., Hochstrasser, R.M.*: J. Amer. Chem. Soc. *98*, 4043 (1976)
46. *Kiel, A., Damen, T.C., Porto, S.P.S., Singh, S., Varsanyi, F.*: IEEE J. Quantum Electron. *4*, 318 (1968)
47. *Griffith, J.S.*: The irreducible tensor method for molecular symmetric groups. London–Tokyo–Sydney–Paris: Prentice Hall International 1962
48. *ibid.* Appendix C
49. *Warshel, A.*: Chem. Phys. Letters *43*, 273 (1976)
50. *Bosworth, Y.M., Clark, R.J.H.*: J. Chem. Soc. (Dalton), 381 (1975)
51. *Gray, H.B.*, in: Transition metal chemistry, Vol. 1, p. 239. Carlin, R.L. (ed.). New York: Marcel Dekker 1965
52. *Clark, R.J.H., Stewart, B.*: to be published
53. *Fano, U., Racah, G.*: Irreducible tensorial sets. New York: Academic Press 1959
54. *Sushchinskii, M.M.*: Raman spectra of molecules and crystals. New York: Keter 1972
55. *McClain, W.M.*: J. Chem. Phys. *55*, 2789 (1971)
56. *Horvath, L.I., McCaffery, A.J.*: J. Chem. Soc., Faraday Trans. 2, *73*, 562 (1977)
57. *Mortensen, O.S.*: Chem. Phys. Letters *3*, 4 (1969)
58. *Clark, R.J.H., Franks, M.L., Turtle, P.C.*: J. Amer. Chem. Soc. *99*, 2473 (1977)
59. *Mortensen O.S., Koningstein, T.A.*: J. Chem. Phys. *48*, 3971 (1968)
60. *Child, M.S., Longuet-Higgins, H.C.*: Phil. Trans. Roy. Soc. *A254*, 259 (1961)
61. *Hamaguchi, H., Harada, I. Shimanouchi, T.*: Chem. Phys. Letters *32*, 103 (1975)
62. *Hamaguchi, H., Shimanouchi, T.*: Chem. Phys. Letters *38*, 370 (1976)
63. *Hamaguchi, H.*: J. Chem. Phys. *66*, 5757 (1977)
64. *Kiefer, W.*: Advances in infrared and Raman spectroscopy, Vol. 3, p. 1. Clark, R.J.H. and Hester, R.E. (eds.). London: Heyden 1977
65. *Clark, R.J.H., Turtle, P.C.*: Inorg. Chem., *17*, 2526 (1978)
66. *Grzybowski, J.M., Andrews, L.*: J. Raman Spectrosc. *4*, 99 (1975)
67. *Baierl, P., Kiefer, W.*: J. Raman Spectrosc. *3*, 353 (1975)
68. *Howard, W.F., Andrews, L.*: J. Amer. Chem. Soc. *97*, 2956 (1975)
69. *Holzer, W., Murphy, W.F., Bernstein, H.J.* : J. Mol. Spectrosc. *32*, 13 (1969)
70. *Clark, R.J.H., Franks, M.L.*: Chem. Phys. Letters *34*, 69 (1975)
71. *Clark, R.J.H., Cobbold, D.G.*: Inorg. Chem., *17*, 3169 (1978)
72. *Stein, L., Norris, J.R., Downs, A.J., Minihan, A.R.*: *Chem. Commun.* in press
73. *Bist, H.D., Brand, J.C.D., Vasudev, R.*: J. Mol. Spectrosc. *66*, 399 (1977)
74. *Andrews, L., Spiker, R.C.*: J. Chem. Phys. *59*, 1863 (1973)
75. *Holzer, W., Murphy, W., Bernstein, H.J.*: Chem. Phys. Letters *4*, 641 (1970)
76. *Homborg, H., Preetz, W.*: Spectrochim. Acta *32*A, 709 (1976)
77. *Clark, R.J.H., Trumble, W.*: J. Chem. Soc. (Dalton), 1145 (1976)
78. *Bosworth, Y.M., Clark, R.J.H.*: unpublished work, Jan. (1972)
79. *Clark, R.J.H., Turtle, P.C.*: Chem. Phys. Letters *51*, 265 (1977)
80. *Funato, Y., Kamisuki, T., Ikeda, K.-I., Maeda, S.*: J. Raman Spectrosc. *4*, 315 (1976)
81. *Cotton, F.A.*: Chem. Soc. Rev. *4*, 27 (1975)
82. *Clark, R.J.H., Franks, M.L.*: J. Amer. Chem. Soc. *97*, 2691 (1975)

83. *Clark, R. J. H., Franks, M. L.*: J. Amer. Chem. Soc. *98*, 2763 (1976)
84. *Clark, R. J. H., D'Urso, N. R.*: J. Amer. Chem. Soc. *100*, 3088 (1978)
85. *Nishimura, Y., Hirakawa, A. Y., Tsuboi, M.*: Advances in infrared and Raman spectroscopy, Vol. 5, p. 217. Clark, R. J. H. and Hester, R. E. (eds.). London: Heyden 1978
86. *Shriver, D. F.*: Advances in infrared and Raman spectroscopy, Vol. 6. Clark, R. J. H. and Hester, R. E. (eds.). London: Heyden, in press
87. *San Filippo, J., Grayson, R. L., Sniadoch, H. J.*: Inorg. Chem. *15*, 269 (1976)
88. *Campbell, J. R., Clark, R. J. H.*: Mol. Phys. *36*, 1133 (1978)
89. *Campbell, J. R., Clark, R. J. H., Griffith, W. P., Hall, J.*: to be published
90. *Robin, M. B., Day, P.*: Adva. Inorg. Chem. Radiochem. *10*, 247 (1967)
91. *Clark, R. J. H.*: Ann. N. Y. Acad. Sci. *313*, 672 (1978)
92. *Clark, R. J. H., Franks, M. L., Trumble, W. R.*: Chem. Phys. Letters *41*, 287 (1976)
93. *Clark, R. J. H., Turtle, P. C.*: Inorg. Chem. *17*, 2526 (1978)
94. *Clark, R. J. H., Franks, M. L.*: J. Chem. Soc. (Dalton) 198 (1977)
95. *Interrante, L. V., Browall, K. W.*: Inorg. Chem. *13*, 1162 (1974)
96. *Mingardi, M., Siebrand, W.*: J. Chem. Phys. *62*, 1074 (1975)
97. *Clark, R. J. H., Stewart, B.*: to be published
98. *Spiro, T. G., Loehr, T. M.*: Advances in infrared and Raman spectroscopy, Vol. 1, p. 98. Clark, R. J. H. and Hester, R. E. (Eds.). London: Heyden 1975
99. *Lewis, A.*: Advances in infrared and Raman spectroscopy, Vol. 6. Clark, R. J. H. and Hester, R. E. (Eds.). London: Heyden, in press
100. *Carey, P. R.*: Advances in infrared and Raman spectroscopy, Vol. 6. Clark, R. J. H. and Hester, R. E. (Eds.). London: Heyden, in press
101. *Clark, R. J. H., Turtle, P. C., Strommen, D. P., Streusand, B., Kincaid, J., Nakamoto, K.*: Inorg. Chem. *16*, 84 (1977)
102. *Clark, R. J. H., Turtle, P. C.*: J. Chem. Soc. (Dalton) 2142 (1977)
103. *Clark, R. J. H., Turtle, P. C.*: J. Chem. Soc. (Dalton) in press (1978)
104. *Jeanmaire, D. L., Van Duyne, R. P.*: J. Electroanal. Chem. *66*, 235 (1975)
105. *Hester, R. E.*: Advances in infrared and Raman spectroscopy, Vol. 4, p. 1. Clark, R. J. H. and Hester, R. E. (eds.). London: Heyden 1978
106. *McQuillan, A. J., Hendra, P. J., Fleischmann, M.*: J. Electroanal. Chem. *65*, 933 (1975)
107. *Paul, R. L., McQuillan, A. J., Hendra, P. J., Fleischmann, M.*: J. Electroanal. Chem. *66*, 248 (1975)
108. *Cooney, R. P., Reid, E. S., Hendra, P. J., Fleischmann, M.*: J. Amer. Chem. Soc. *99*, 2002 (1977)
109. *Creighton, J. A.*: unpublished work
110. *Friedman, J. M., Rousseau, D. L.*: Chem. Phys. Letters, *55*, 488 (1978)

Structure and Bonding of Alkali Metal Suboxides

Arndt Simon

Max-Planck-Institut für Festkörperforschung, Heisenbergstr. 1, D-7000 Stuttgart 80

Table of Contents

The alkali metals Rb and Cs form metal-rich oxides (suboxides) with unusual compositions and structures. This article aims at a verification of the simple bond model derived from the crystal structures of these compounds by the measurement of a variety of physical properties. According to this aim the interesting aspects of application of alkali metal suboxides, e.g. in photocathodes, are only discussed very briefly. – The description of the compounds covers their phase relationships and crystal structures. The structural details are analyzed with respect to the different kinds of chemical bonding (ionic and metallic) inside and between characteristic clusters Rb_9O_2 and $Cs_{11}O_3$ occurring in the binary and Rb/Cs-mixed compounds. Calculations show that these clusters are stable configurations of the M^+ and O^{2-} ions in a constraining field of metallic electrons. Raman spectroscopy and electrical conductivity measurements are used to prove both aspects qualitatively. Optical reflectivity measurements and especially photoelectron spectroscopy provide a quantitative verification of the bond model. – Somewhat off the main line of the article recent results on phase formation and stability including investigations on glass-like metallic states are referred. The easy achievement of such metastable states makes the alkali metal suboxides very interesting systems to study these phases and the equilibration processes.

I. Introduction

It has been and still is a challenge to chemists to prepare compounds of metals in high oxidation states. But in the last twenty years in particular there has been a growing interest in compounds containing metals in unusually low formal oxidation states. Such a condition leaves valence electrons with the metal atoms capable of forming metal-metal bonds. Metal-metal bonding widely occurs in compounds with transition metals (*1–4*), some main group elements (*5*), and to a growing extent, even with the lanthanoids. (*6, 7*) The existence of discrete metal atom clusters or infinitely extending regions with homonuclear metal bonds – often with low dimensionality – is of interest to chemists as well as physicists.

It is a surprising fact that alkali metals also occur in low formal oxidation states: Suboxides like Cs_7O and Cs_4O have been known to exist for a long time. (*8, 9*) During the last ten years these suboxides have been investigated extensively; (*10*) this investigation has brought to light the existence of a very unique class of compounds. Looking at the alkali metal suboxides in a simplified way, their structural chemistry is somewhat inverse to that of metal-rich transition metal compounds. The latter (such as cluster compounds of elements from groups VB, VIB and VIIB) reveal strong directional and thus rigid metal-metal bonds dominating the crystal structures. In contrast to that, the bonding between alkali metal atoms is non-directional and "soft". So the structural chemistry of the metal-rich alkali metal oxides is dominated by the strong hetero atom interactions.

Preparative aspects of the investigations of alkali metal suboxides, e.g. synthesis, analysis and crystal growth have been described in detail earlier. (*11, 12*) The emphasis of this paper lies on structural aspects, measurements of some important physical properties and their discussion in terms of the chemical bonding within these compounds. Some attention is paid to the structure and bonding in metastable and amorphous states of alkali metal suboxides, too.

II. Compounds and Phase Relations

1. *Binary Compounds*

In the Rb/Rb_9O_2 system the two stable compounds Rb_6O and Rb_9O_2 exist (*11*) as shown in Fig. 1a. The compound Rb_9O_2, which is identical with the compound analyzed as "Rb_3O" by *Touzain*, (*13, 14*) decomposes peritectically at about 308 K (*15*) to form "normal" salt-like Rb_2O. A eutectic system melting at 266 K is formed between Rb and Rb_9O_2. Only 0.3 K below this temperature the compound Rb_6O decomposes. (*16*) Due to kinetic reasons both heterogeneous conditions (Rb_6O + Rb as well as Rb_9O_2 + Rb) can be achieved nearly across the entire range of composition by applying different cooling rates to the melts. The different results are closely related to the existence of amorphous and metastable crystalline phases below the dashed horizontal line: Rapid quenching of a melt Rb_xO ($x > 6.5$) leads to a glass-like state besides crystalline Rb. According to X-ray results (*17*) and measurements of the electrical conductivity (*18*) the amorphous portion seems to have a constant composition of approximately $RbO_{0.167}$ ($\sim Rb_6O$). The range, where such a quenched sample is nearly all amorphous, is indicated in Fig. 1a. When these samples are slowly warmed, intermediate unstable crystalline compounds such as $Rb_{6.33}O$ form at approximately 140 K, (*17*) and decompose at about 180 K. The decomposition always leads to the mixture Rb_9O_2 + Rb, which is a metastable system at this temperature, too. In contrast, slow cooling of a melt results in the preferred crystallization of Rb_6O (+ Rb).

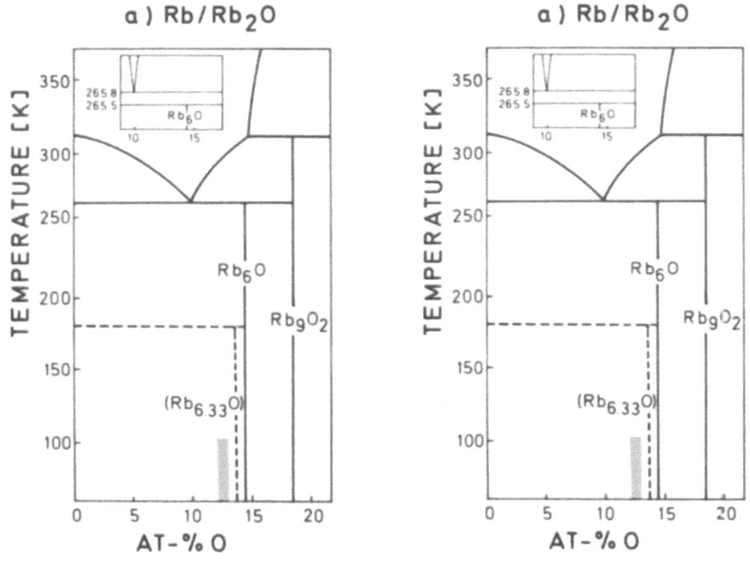

Fig. 1. Phase diagrams (*11, 17, 18*) of the systems (a) $Rb-Rb_2O$ and (b) $Cs-Cs_2O$. Dotted lines refer to nonequilibrium conditions with the stable compounds. Metastable crystalline phases, e.g. $Rb_{6.33}O$, exist in the $Rb-Rb_2O$ system. The shaded areas indicate regions where metallic glasses are obtained by quenching the melts

Needless to say, such crystallization is also a non-equilibrium process. It is this complicated kinetic behavior that renders physical measurements on well-defined samples of Rb suboxides (as well as Cs suboxides) very difficult.

The corresponding results gained in the Cs—Cs_2O system are summarized in Fig. 1b. The stoichiometric compounds Cs_7O, Cs_4O and $Cs_{11}O_3$ are present in addition to a phase "Cs_3O", which exhibits a broad range of homogeneity ($CsO_{0.31-0.36}$). *(19)* The crystallization of the stable phases in the Cs—Cs_2O system is delayed very much as with the Rb suboxides. This is indicated by dotted lines in the diagram. E.g. the compound Cs_4O does not form readily; indeed, with small samples (some milligram) this compound could never be observed, obviously because of a low probability of nuclei formation.

Quenching liquid samples of the composition Cs_xO ($4 < x < 6$) leads to glass-like metals. *(20)* According to X-ray investigations, these samples are entirely amorphous in contrast to the Rb—O samples, which always contain variable amounts of crystalline Rb and suboxides, respectively. Crystallization occurs at about 140 K. In samples which contain both alkali metals Rb and Cs, as well as O, the tendency to form the glass-like state is even more enhanced.

As pointed out in Section V.2., the amorphous alkali metal systems might be of considerable interest with respect to the understanding of metallic glasses.

2. *Ternary Compounds*

One might ask whether the formation of solid stable suboxides with main-group metals is confined to the heavy alkali metals Rb and Cs. Several attempts have been made to prepare such compounds with the lighter alkali metals. According to these experiments, Li and Na are unable to form binary or ternary suboxides. Partial oxidation of low melting alloys like K—Na always results in the deposition of the "normal" oxide, e.g. Na_2O. These results are understood as being due to the high stability (large values of lattice energy) of these oxides.

Ternary suboxides are formed by oxidizing a K—Cs alloy, but these decompose at 210 K. From the knowledge of Rb—Cs suboxides it can be concluded that the K/Cs suboxides are of the type $(K, Cs)_nCs_{11}O_3$ (see III.1., 2.). A detailed investigation in the K—Cs—O system has not been undertaken because of the extreme experimental difficulties.

Even in the Rb—Cs—O system the difficulties in preparing equilibrium phases have hardly been overcome. Some preliminary results are illustrated in Fig. 2. *(21)* The ternary phase diagram along the line Rb—$Cs_{11}O_3$ closely resembles the binary system between Cs and $Cs_{11}O_3$. The close relationship becomes even more evident when discussing the structures (see III.1.). $Cs_{11}O_3Rb$, *(22)* $Cs_{11}O_3Rb_2$ *(22)* and $Cs_{11}O_3Rb_7$ *(21)* are the only compounds which could be characterized analytically and structurally until now. But there is still some ambiguity in the discussion of experimental results with respect to the equilibrium phase diagram. The difficulties are due to the pronounced preference for non-equilibrium states as well as the occurrence of ranges of homogeneity.

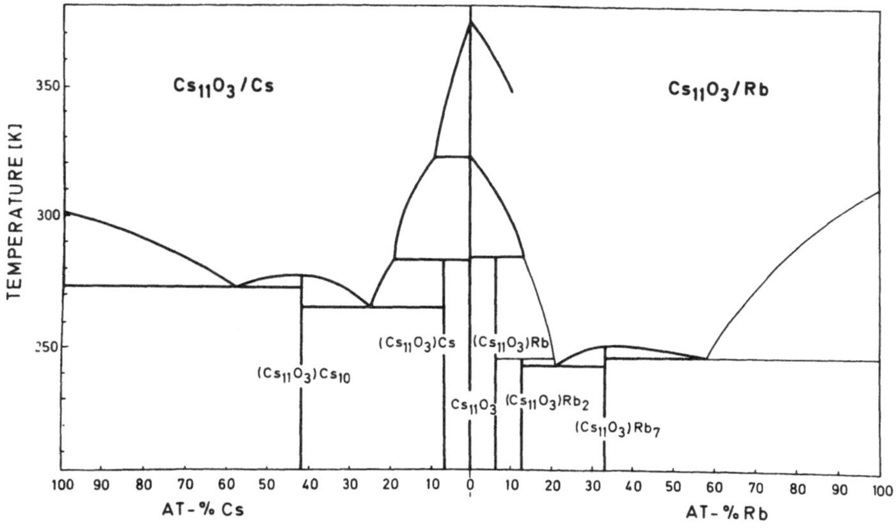

Fig. 2. Comparison of the partial phase diagrams $Cs_{11}O_3$–Cs and $Cs_{11}O_3$–Rb. *(21, 22)* The phase relations described in the oxygen-rich part of the $Cs_{11}O_3$–Rb diagram (between $Cs_{11}O_3$ and $Cs_{11}O_3Rb_2$) are still preliminary. The abscissa values refer to the content of B-type atoms. See page 9

III. Structures of Alkali Metal Suboxides

1. Structures of Stable Compounds

The crystal structures of all binary suboxides (with the exception of "Cs_3O") as well as the structures of $Cs_{11}O_3Rb$, $Cs_{11}O_3Rb_2$ and $Cs_{11}O_3Rb_7$ have been solved by X-ray single crystal methods. The crystallographic data are summarized in Table 1. All structures follow a unique, but uniform principle:

(a) The O atoms are surrounded by octahedra of Rb or Cs atoms.
(b) Face sharing of two such octahedra results in the cluster Rb_9O_2; three equivalent octahedra form the cluster $Cs_{11}O_3$.
(c) O–M distances are near the values expected for M^+ and O^{2-} ions. The ionic character of the M atoms is expressed in short intra-cluster M–M distances.
(d) The inter-cluster M–M distances are comparable to the distances in Rb and Cs.
(e) The clusters and additional alkali metal atoms form compounds of new stoichiometries.

Table 1. Crystallographic data of alkali metal suboxides; z and V refer to the structural formula given first. Temperatures are added in brackets. The thermal expansion $\alpha = \frac{1}{v}\frac{dv}{dT}$ lies between $100 \cdot 10^{-6}\ K^{-1}$ for Rb_9O_2 and $Cs_{11}O_3$ and $200 \cdot 10^{-6}\ K^{-1}$ for Rb and Cs

Formula	Space group	Unit cell [pm] at (T [K])	V per formula unit [$cm^3 \cdot mol^{-1}$] at (T [K])	Remarks	Ref.
Rb_9O_2	$P2_1/m$, z = 2	832.3, 1398.6, 1165.4, 104.39° (223)	395.8 (223)	Accurate lattice constants from Guinier photographs	*(15)*
$Rb_9O_2Rb_3$ (≙ Rb_6O)	$P6_3/m$, z = 2	839.3, 3046.7 (223)	559.7 (223)	Same as with Rb_9O_2	*(16)*
$Cs_{11}O_3$	$P2_1/c$, z = 4	1753.6, 919.2, 2401.7, 100.42° (223)	573.4 (223)	Same as with Rb_9O_2	*(10, 24)*
$Cs_{11}O_3Cs$ (≙ Cs_4O)	Pna2, z = 4	1682.3, 2052.5, 1237.2 (173)	643.3 (173)	Lattice constants with large standard deviations from Guinier photographs	*(28)*
$Cs_{11}O_3Cs_{10}$ (≙ Cs_7O)	$P\bar{6}m2$, z = 1	1630.9, 915.4 (223)	1269.9 (223)	Same as with Rb_9O_2	*(26)*
$Cs_{11}O_3Rb$	$Pmn2_1$, z = 2	1646.4, 1368.3, 909.2 (243)	616.9 (243)	Diffractometer data, Rb/Cs ratio not known analytically	*(22)*
$Cs_{11}O_3Rb_2$ (I)	$P2_1$, z = 8	4231.1, 919.4, 2415.6, 108.94° (223)	669 (223)	Same as $Cs_{11}O_3Rb$	*(22)*
(II)	orthorhomb., z = 2	2028, 1225, 902.3 (183)	675 (183)	Same as $Cs_{11}O_3Rb$	*(32)*
$Cs_{11}O_3Rb_7$	$P2_12_12_1$, z = 4	3228.1, 2187.7, 902.5 (183)	959.9 (183)	Same as Rb_9O_2	*(21)*

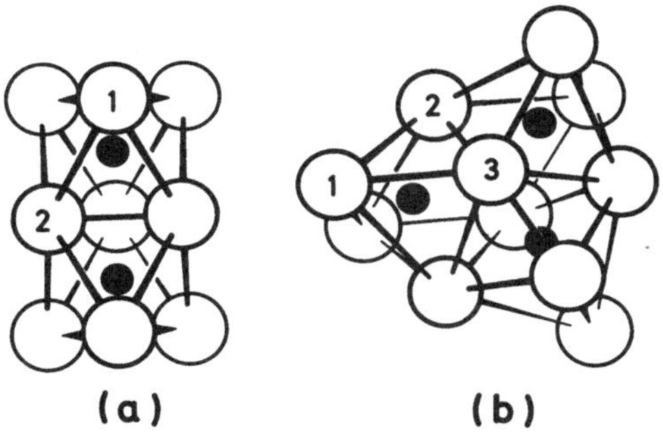

Fig. 3. Rb_9O_2 (a) and $Cs_{11}O_3$ (b) cluster (open circles: Rb, Cs; black circles: O). The metal atoms are numbered according to their c.n. towards oxygen. For interatomic distances refer to Table 2

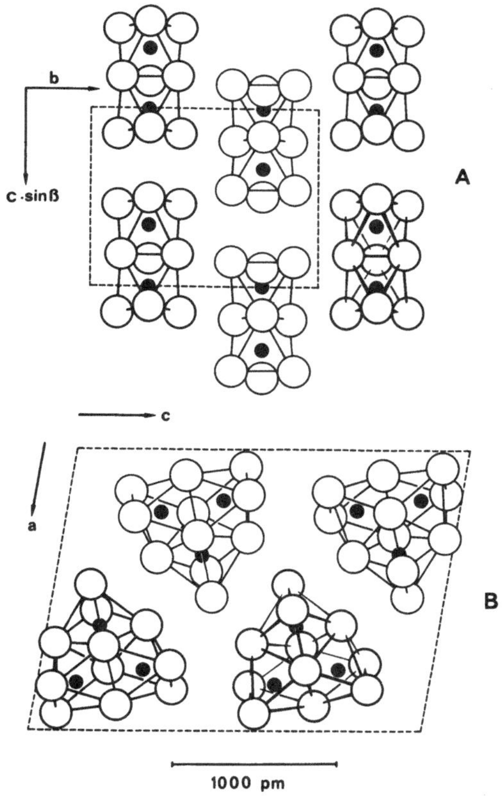

Fig. 4. Projections of the cluster arrangements in (A) Rb_9O_2 *(15)* and (B) $Cs_{11}O_3$ *(24)* (open circles: Rb, Cs; black circles: O)

It should be mentioned that the *anti*-types of the clusters are well known: The complex halide anion $M_2X_9^{3-}$ occurs with many main group and transition metals. The M_3X_{11} cluster occurs (but not as an isolated unit) in Nb_3Se_4. *(23)* The characteristic structural units of alkali metal suboxides are shown in Fig. 3. The Rb_9O_2 clusters are only found in pure Rb suboxides and highly oxidized samples (see III.2.); the $Cs_{11}O_3$ cluster is a constituent of the Cs and Rb–Cs suboxides.

In Fig. 4 projections of the monoclinic unit cells of Rb_9O_2 *(15)* and $Cs_{11}O_3$ *(24)* which are characteristic of the clusters, are chosen. Both structures are essentially "molecular crystals" consisting of Rb_9O_2 and $Cs_{11}O_3$ clusters only. There is a fundamental difference between the two clusters: In contrast to the $Cs_{11}O_3$ cluster, the Rb_9O_2 cluster exhibits an arrangement of metal atoms, which is part of a (hexagonal) close packing. The analysis of the atom contacts between adjacent clusters therefore is simple with the Rb-suboxides: The clusters are packed such that an infinitely close-packed array of Rb atoms results. The close packing principle is also stressed by the very special kind of cluster arrangement in Rb_9O_2, which reveals a delicate balance of distances with respect to the different atomic distances between equatorial and peripheral Rb atoms in the cluster (see Table 2).

Table 2. Interatomic distances [pm] between neighboring atoms in alkali metal suboxides. The distances in the clusters are mean values and characterized by the addition of the coordination number (towards O) to M (see Fig. 3). Distances between clusters and B-type atoms (M–M) are characterized by the smallest occurring values

Cluster	$Cs_{11}O_3$					Rb_9O_2	
compound	$Cs_{11}O_3$	Cs_4O	Cs_7O	$(Cs_{11}O_3)Rb_7$	$Cs_{11}O_3Rb$	Rb_9O_2	Rb_6O
M1–M1	406	406	402	402	404	394	394
M1–M2	431	430	427	427	427	403	400
M1–M3	416	416	414	414	413	–	–
M2–M2	–	–	–	–	–	354	354
M2–M3	377	376	375	374	373	–	–
M3–M3	367	368	371	371	360	–	–
M1–O	275	270	273	273	266	270	270
M2–O	289	293	289	290	293	282	280
M3–O	292	300	295	296	299	–	–
O–O	404	406	392	–	409	391	382
M–M	479	489	518	471	498	474	487
	520	506	532	496	501	500	494
	527	515	538	498	510	507	519

In the $Cs_{11}O_3$ cluster the Cs atoms do not conform to any infinite close-packed array. Yet the arrangement of clusters in the compound $Cs_{11}O_3$ (and the other Cs and Rb–Cs suboxides) is traced back to very special conditions for the Cs–Cs contacts between adjacent clusters. Due to the low-symmetry environment of the cluster in $Cs_{11}O_3$, interatomic distances of equivalent kinds differ notably (Table 2). But the large range of distances between adjacent atoms of neighboring clusters reveals the comparably rigid character of this unit.

When the clusters in Rb_9O_2 and $Cs_{11}O_3$ are idealized by spheres A ($\hat{=}$ $Cs_{11}O_3$ cluster) and A' ($\hat{=}$ Rb_9O_2 cluster), the structures can easily be described as close-packed arrangements of such spheres. The large "pseudo-atoms" A and A' combine with additional alkali metal Rb and Cs ($\hat{=}$ B) to form compounds of stoichiometry AB_{10}, AB_7, $A'B_3$, AB_2 and AB.

In Fig. 5 this principle is illustrated for the compound Rb_6O (*16, 17*), which, according to its structure, has to be written as $Rb_9O_2Rb_3$ ($\hat{=}$ $A'B_3$). Layers of Rb_9O_2 clusters alternate with close-packed layers of pure Rb. The structure is related to that of $BaTiO_3$ ($\hat{=}$ $Ba_3TiTi_2O_9$) by leaving one octahedral site unoccupied ($\hat{=}$ $Ba_3\square Ti_2O_9$). (*25*) The size of the empty interstitial site is reduced by the rotation of the clusters around their three-fold axis. As with Rb_9O_2 the distances between atoms of adjacent clusters are nearly the same as in Rb. The B-type atoms, as inferred from interatomic distances, exhibit metallic bonding to all surrounding atoms.

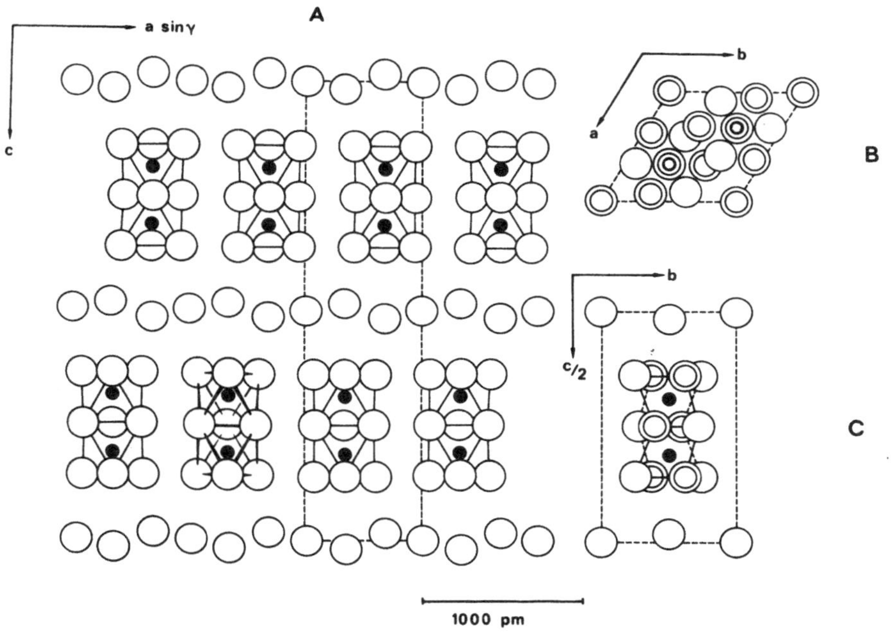

Fig. 5. Projections of the hexagonal ordered (A, B) structures of Rb_6O. (*16*) (C) is a projection of the disordered structure found with rapidly crystallized samples (*17*) (atoms denoted as in Fig. 4)

As one might guess, the crystal structure of Rb_6O is strongly influenced by crystallographic disorder of the layers. The X-ray pattern of rapidly crystallized Rb_6O corresponds to the twin model (*17*) shown in Fig. 5c. A rotational disorder within each Rb_9O_2 layer can definitely be excluded. On the contrary, there is evidence for a very high degree of order within each Rb_9O_2 layer. As the Rb atoms of the clusters are close-packed, the single cluster is hindered from achieving the right rotational position for the ordered crystal in spite of the high degree of thermal motion of the Rb atoms (see IV.3.). So we find the interesting phenomenon that ordering of Rb_6O has to proceed *via* a collective rotation (by 60°) of the clusters belonging to one layer. According to this mechanism, ordering does not occur completely until a few tenths of a degree below the decomposition temperature of Rb_6O at 265.5 K.

The structure of Rb_6O suggests the possibility for a series of Rb suboxides to exist with higher metal contents by inserting, for instance, more than one layer of B atoms between the layers of clusters. Although evidence for the existence of further (stable) Rb suboxides is given by thermal investigations, (*11*) these have never been observed with X-ray investigations.

Fortunately, a number of suboxides which contain the $Cs_{11}O_3$ cluster exist. The hexagonal structure of Cs_7O (*26*) is shown projected along [001] in Fig. 6A. It is described as $Cs_{11}O_3Cs_{10}$ ($\triangleq AB_{10}$). In the B-atom environment of exactly trigonal symmetry the $Cs_{11}O_3$ clusters are undistorted and form a chain of coplanar units along their three-fold axis. The B-type Cs atoms occupy certain geometrically preferred positions around this chain, such that all triangular faces of the $Cs_{11}O_3$ cluster are centered by B-type atoms:
Three Cs atoms (nearest to the three-fold axis) are attached to four inner triangular faces of two adjacent clusters, filling the "valleys" on the surface of the chain.
Six Cs atoms are attached to the external triangular faces of one cluster.
Three Cs atoms are attached to edges of the clusters and six atoms complete the cluster surrounding in such a way that the arrangement of B atoms between the chains is close-packed.

It should be noted that each B-type atom coordinates 1, 2 and 3 cluster chains; there is no isolated Cs atom in this matrix of Cs atoms. As will be pointed out later, the principal features of the characteristic environment of the $Cs_{11}O_3$ cluster in Cs_7O are found in every compound containing this cluster in spite of different stoichiometries and cluster arrangements.

The very similar structure principle is obvious with $Cs_{11}O_3Rb_7$ ($\triangleq AB_7$), (*21*) the structure of which is drawn as a projection along [001] in Fig. 6B. The compound serves as a very surprising example of a strikingly different chemical behavior for Rb and Cs: Simply oxidizing an alloy of Rb and Cs leads to a spatial separation of the two metals in $Cs_{11}O_3Rb_7$. Following the refinement of occupation factors, this separation is found to be complete. In spite of the different stoichiometries the structures of Cs_7O and $Cs_{11}O_3Rb_7$ are very similar. The $Cs_{11}O_3$ clusters are only slightly inclined within the chain in the structure of $Cs_{11}O_3Rb_7$. The three Rb atoms nearest to the quasi-trigonal axis have moved into the "valleys" and the identity period of the chain has become slightly smaller with $Cs_{11}O_3Rb_7$ due to the smaller size of these

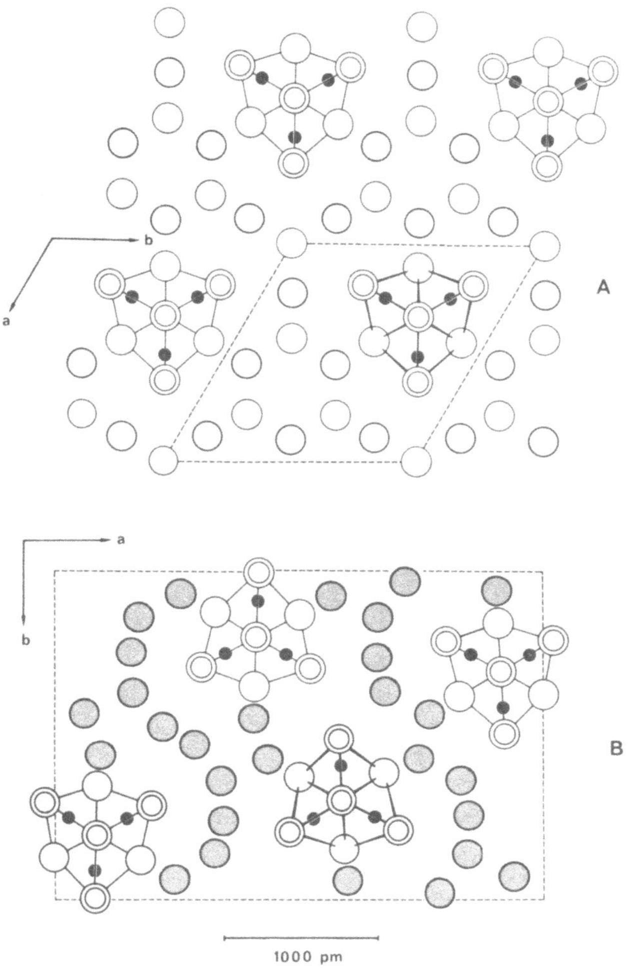

Fig. 6. Projections of the hexagonal structure of Cs_7O (*26*) (A) and the orthorhombic unit cell of $Cs_{11}O_3Rb_7$ (*21*) along [001] (B). Open circles symbolize Cs, shaded circles Rb and black circles O atoms

atoms. The other positions around the chains are occupied in very much the same way as in Cs_7O but now B-type Rb atoms and Cs atoms of adjacent clusters belong to the cluster environment.

The structure (*22*) of $Cs_{11}O_3Rb_2$ ($\triangleq AB_2$), shown in Fig. 7 conforms to the geometrical arrangement around the cluster chains, although the number of B-type atoms is not sufficient to occupy the characteristic positions mentioned with Cs_7O: There are two types of cluster chains in $Cs_{11}O_3Rb_2$; one is surrounded by three inner Rb atoms as in Cs_7O and $Cs_{11}O_3Rb_7$, whereas the other is combined with its equiva-

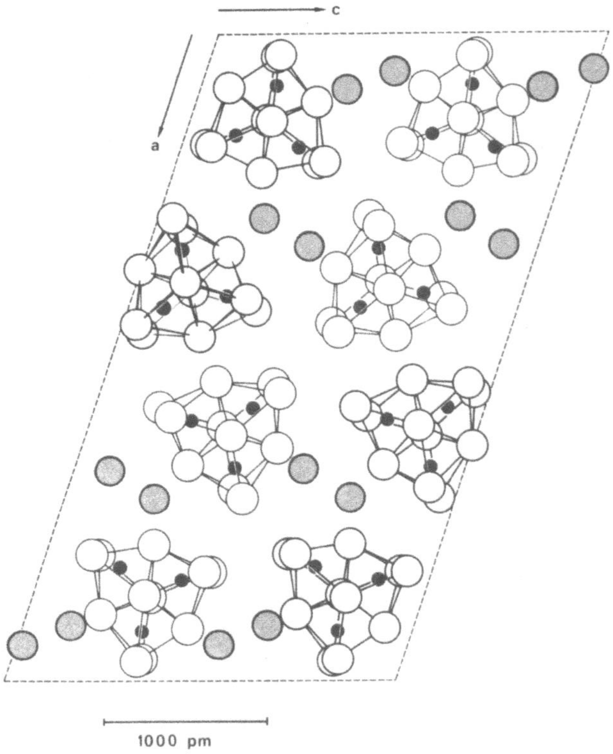

Fig. 7. Projection of the $Cs_{11}O_3Rb_2$ structure *(22)* along [010]. The atoms are denoted as in Fig. 6

lent to form a double chain. In this double chain one of the equatorial Cs atoms of the cluster mutually takes the positions of such inner B-type atom; the remaining four "valleys" of the double chains are occupied by Rb atoms in the known manner. There is no even distribution of Rb atoms in the structure of $Cs_{11}O_3Rb_2$.

According to its structure the compound is an intergrowth of Rb-rich and -poor components. In addition to the discussed structure a polymorphic form of $Cs_{11}O_3Rb_2$ exists, which is still under investigation. *(27)* The structure *(22)* of $Cs_{11}O_3Rb$, shown in Fig. 8A, corresponds to that of $Cs_{11}O_3Rb_2$. The association of cluster chains has proceeded into waved layers along the (001) plane. Two thirds of the "valley" positions of each chain are occupied by Cs atoms of adjacent chains; the Rb atoms are attached to the remaining positions. The close relationship to the structure of $Cs_{11}O_3$ is evident. Indeed, $Cs_{11}O_3Rb$ results from an occupation of interstitial sites in $Cs_{11}O_3$ by Rb atoms, accompanied by a very moderate reorientation of the clusters. The idealized arrangement of clusters and Rb atoms corresponds to the WC structure (W ≙ cluster, C ≙ Rb).

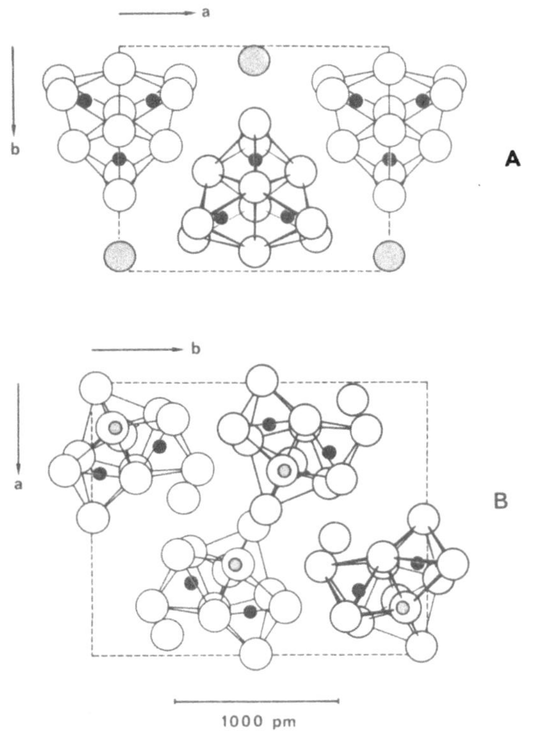

Fig. 8. Projection of the $Cs_{11}O_3Rb$ structure (*22*) (A) and the structure of Cs_4O (*28*) (B) along [001]. The atoms are denoted as in Fig. 6

Obviously, the Rb site in $Cs_{11}O_3Rb$ is too small for Cs. The structure (*28*) of $Cs_4O \triangleq Cs_{11}O_3Cs$ ($\triangleq$ AB) differs markedly (Fig. 8B), in that the clusters are strongly inclined to one another. The best way of describing the overall arrangement of clusters and single atoms in terms of an AB structure is to point out the relationship to the NiAs type structure.

As far as suboxides containing cesium with a ratio $O/M \leqslant 3/11$ are concerned, it has been shown that characteristic $Cs_{11}O_3$ clusters occur along with additional alkali metal atoms. It is an interesting question, therefore, whether the same clusters are also the main constituents of the phase "Cs_3O". Such a hypothesis leads to a modified structure principle, where $Cs_{11}O_3$ clusters alternate with intermediate regions of Cs_2O instead of Cs atoms.

"Cs_3O" has been described as crystallizing in the *anti*-TiI_3 structure type. (*29*) Although this structure model has now (*30*) been refined on the basis of diffractometer data down to R = 0,044, the model definitely does not account for all observations. Thus the temperature factors remain unusually large even at 130 K. Strong

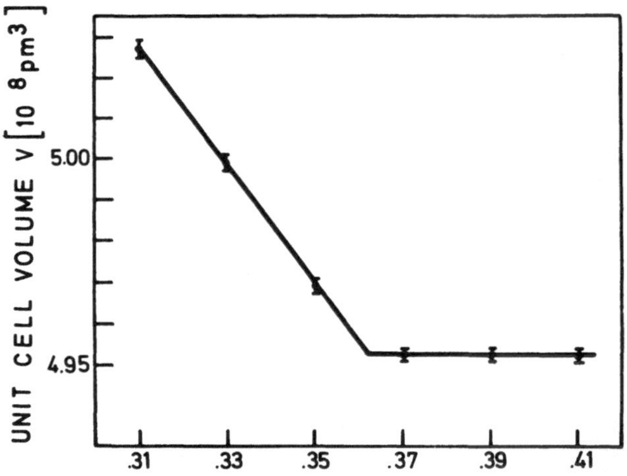

Fig. 9. Variation of the unit cell volume with sample compositions for the homogeneous phase "Cs_3O" at room temperature (*19*) (modified Guinier technique). Error bars indicate the standard deviations of the measurements

diffuse scattering effects occur with different crystals indicating pronounced disorder to be typical with "Cs_3O". Moreover, "Cs_3O" exhibits a range of homogeneity (Fig. 9) (*19*), which extends to both sides of the ideal composition $CsO_{0.33}$ and which is difficult to understand on the basis of a TiJ_3 type structure. Unfortunately, the growth of suitable single crystals with different (and exactly known) compositions has not been achieved so far to allow us to find an appropriate disorder model.

2. *Metal Substitution Reactions*

The chemically different behavior of Rb and Cs is demonstrated in the structures of the Rb—Cs mixed suboxides. It is interesting to learn about the chemical similarity, too, by investigating, whether a mutual exchange of these metals is possible. Experiments have led to initial answers to the questions, which of the clusters, Rb_9O_2 or $Cs_{11}O_3$, is the more stable one and what the conditions are to substitute the metal atoms specifically in the clusters and in the oxygen-free parts of the structures.

The reaction of $Cs_{11}O_3$ with pure Rb leads to compounds $Cs_{11}O_3Rb_x$. Some of these have been characterized. Reaction of Rb_9O_2 with an excess of Cs, however, results in the decomposition of Rb_9O_2 clusters and formation of the $Cs_{11}O_3$ cluster. Compounds of the type $Rb_9O_2Cs_x$ do not exist. Consequently, the oxidation of Rb—Cs alloys yields compounds with pure $Cs_{11}O_3$ clusters as long as there is a ratio Cs–O $\geqslant 11/3$ in the sample. In the graphical representation of Fig. 10 this means that the occurrence of the pure $Cs_{11}O_3$ cluster is confined to compounds which lie in the

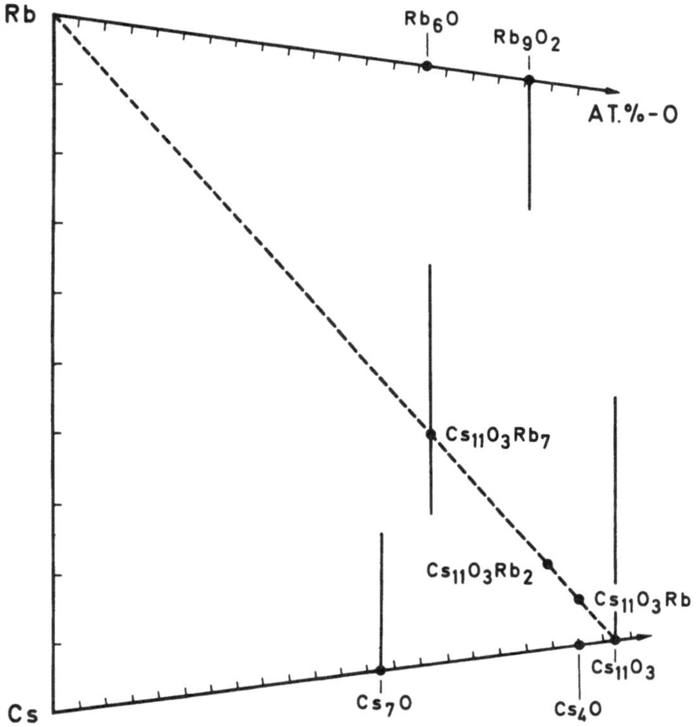

Fig. 10. Ranges of homogeneity in the ternary phase diagram of Rb–Cs suboxides. *(21, 30)* The vertical lines in the diagram correspond to the deviations from exact stoichiometry found experimentally

compositional triangle Rb—Cs—$Cs_{11}O_3$. According to X-ray analyses a metal exchange in these compounds is only possible with B-type atoms leading to compounds of the general formula $Cs_{11}O_3(Rb, Cs)_x$. *(21)* As indicated in Fig. 10, the ranges of homogeneity strictly follow the lines of constant O/M ratios, except for unmeasurably small deviations. In Fig. 11 the results of exchange reactions with Cs_7O are summarized. *(31)* It is possible to substitute up to about 20 at-% of the Cs atoms by Rb atoms. The exchange is accompanied by a drastic decrease in the melting point of Cs_7O and a decrease in the lattice constants. By considering the volume changes, the reaction is unambiguously interpreted as a substitution within the purely metallic parts in Cs_7O according to the formula $Cs_{11}O_3(Cs, Rb)_{10}$. This conclusion can be drawn on the basis of the values of Table 1.

The molar volumes of the metal-rich suboxides are nearly equal to the sum of the cluster volumes taken from the compounds Rb_9O_2 and $Cs_{11}O_3$ (396 and 573.5 $cm^3 \cdot mol^{-1}$, respectively) and the atomic volumes of the additional metal taken from the

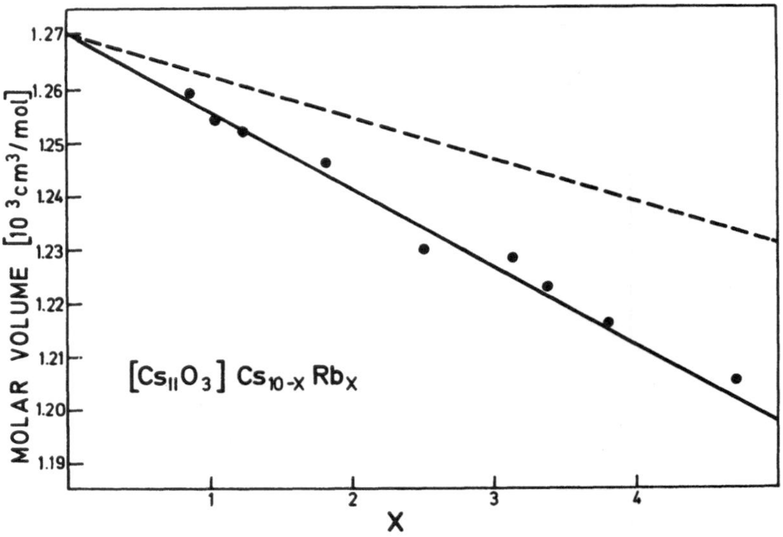

Fig. 11. Change of molar volume of Cs_7O, when part of the Cs atoms is substituted by Rb atoms at 223 K. (*31*) The full line is calculated for a substitution reaction according to $Cs_{11}O_3$ (Rb, Cs)$_{10}$. The dashed line corresponds to a reaction according to (Cs, Rb)$_nO_3Cs_{10}$

elements Rb and Cs (54.7 and 69.2 $cm^3 \cdot mol^{-1}$, respectively). This result is in accordance with the metallic bonding of the B-type atoms to their environment. Therefore, the substitution of one Cs atom of B-type by an Rb atom leads to a volume decrease of 14.5 $cm^3 \cdot mol^{-1}$. Substitution of cluster atoms should result in a much smaller decrease of the volume. This follows from the fact that the mean volume increment of Rb and Cs in the cluster is approximately 41.5 and 49 $cm^3 \cdot mol^{-1}$, respectively, taking into account an increment of about 11 $cm^3 \cdot mol^{-1}$ for O^{2-}. (*33*) Thus a volume change of 7.5 $cm^3 \cdot mol^{-1}$ for one exchanged cluster atom is expected. The full line in Fig. 11 is constructed with $\Delta V = 14.5$, the dashed line with $\Delta V =$ 7.5 $cm^3 \cdot mol^{-1}$. Up to about five Cs atoms in $Cs_{11}O_3Cs_{10}$ can be substituted by Rb atoms. Unfortunately, suitable single crystals of the substituted compound are not yet available to investigate whether the Rb atoms occupy preferred positions within the set of B atoms.

Beyond the borderline $Cs_{11}O_3$–Rb towards higher oxidation state all Cs is incorporated in clusters and thus coordinated to O atoms. Under this condition substitution of the Rb_9O_2 cluster by Cs atoms as well as substitution of Cs atoms by Rb atoms in the $Cs_{11}O_3$ cluster is observed.

The Rb_9O_2 cluster substituted by Cs atoms was found with a sample of arbitrary composition $(Rb_9O_2)_{.76}(Cs_{11}O_3)_{.24}$ ($\hat{=} Rb_{6.1}Cs_{2.4}O_2$). The X-ray pattern of this sample is indexed essentially on the basis of an enlarged unit cell of Rb_9O_2. Instead of a molar volume of 396 $cm^3 \cdot mol^{-1}$ a value of 413 $cm^3 \cdot mol^{-1}$ is found for the substituted Rb_9O_2. The additional 17 $cm^3 \cdot mol^{-1}$ correspond to a substitution of

2 to 3 Rb atoms by Cs in the clusters. Such metal substitution reactions have not yet been studied in detail with the Rb_9O_2 cluster, but some structural work has been focussed on the $Cs_{11}O_3$ cluster.

From a melt of composition $Cs_5Rb_6O_3$ single crystals can be grown, which have the structure of $Cs_{11}O_3$ but with smaller lattice constants. The cluster in this phase has a molar volume of 527.8 instead of 573.5 $cm^3 \cdot mol^{-1}$ for $Cs_{11}O_3$ itself. The difference of approximately 46 $cm^3 \cdot mol^{-1}$ is in very good agreement with the expected value for the substitution of six Cs atoms by Rb. Refinement of the occupation factors yields an interesting ordering of the different kinds of alkali metal atoms in the substituted cluster (*22*) (Fig. 12): The central positions (3) are entirely occupied by Rb atoms as well as one position of kind (1). All other atoms are on the average exchanged by 50% (atoms 1) and 65% (atoms 2), refined relative to the weight of the oxygen atoms. Hence the absolute values of the occupation factors are poorly scaled.

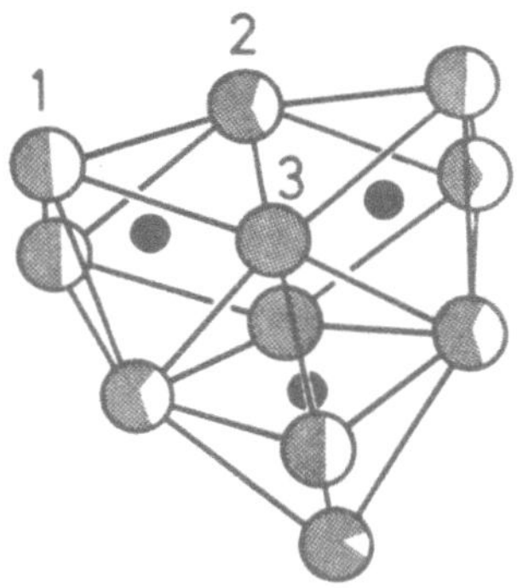

Fig. 12. Substitution of Cs atoms by Rb atoms in $Cs_{11}O_3$. The amount of replacement is proportional to the shaded areas (*22*)

The selective exchange of Cs by Rb atoms most probably reflects general aspects concerning the bonding in the cluster. Thus the substitution of atoms (3) with c.n. = 3 towards oxygen by smaller atoms gains a maximum in Coulomb energy due to a reduction of (6) M–O distances from 296 to 286 pm. On the other hand, the exchange reflects specific aspects concerning the $Cs_{11}O_3$ structure, as the one entirely exchanged Cs 1 atom has the shortest known distance of 479 pm to an atom of the adjacent cluster in $Cs_{11}O_3$. Thus the substitution by the smaller Rb atom is favored via the special cluster arrangement.

Last, but not least, the compounds $Cs_{11}O_3Rb_x$, which are located on the borderline between the observed substitution reaction in the clusters and in the O-free parts of the structure, have to be considered. Due to the difficulties in the establishment of phase equilibria, investigations with these compounds are in a very preliminary state. Yet it has been shown that $Cs_{11}O_3Rb_7$ exhibits a broad range of homogeneity along the line of constant oxidation number within the approximate limits $Cs_{13}Rb_5O_3$

and $Cs_6Rb_{12}O_3$. (21) As a consequence, the stoichiometric phase $Cs_{11}O_3Rb_7$ itself has to be formulated as $Cs_{11-x}Rb_xO_{7-x}Cs_x$, but according to the distances and occupation probabilities, there is no experimental evidence for a mutual replacement of atoms in the phase of exact composition ARb_7.

3. Structures of Liquid and Metastable Suboxides

Complex ionic clusters are the constituents of the stable alkali metal suboxides. One also might expect the same clusters to occur in melts of these compounds.

In Fig. 13 the neutron scattering result of a measurement on liquid "Rb_6O" is shown. (34) S(Q) is characterized by a pronounced decrease in the intensity of the main peak accompanied by a shift from 1.52 to 1.58 $Å^{-1}$. An additional peak arises at 68 $Å^{-1}$, which corresponds to pair correlations at long distances around 920 pm in real space. Characteristic features of the ionic clusters are recognized in the pair

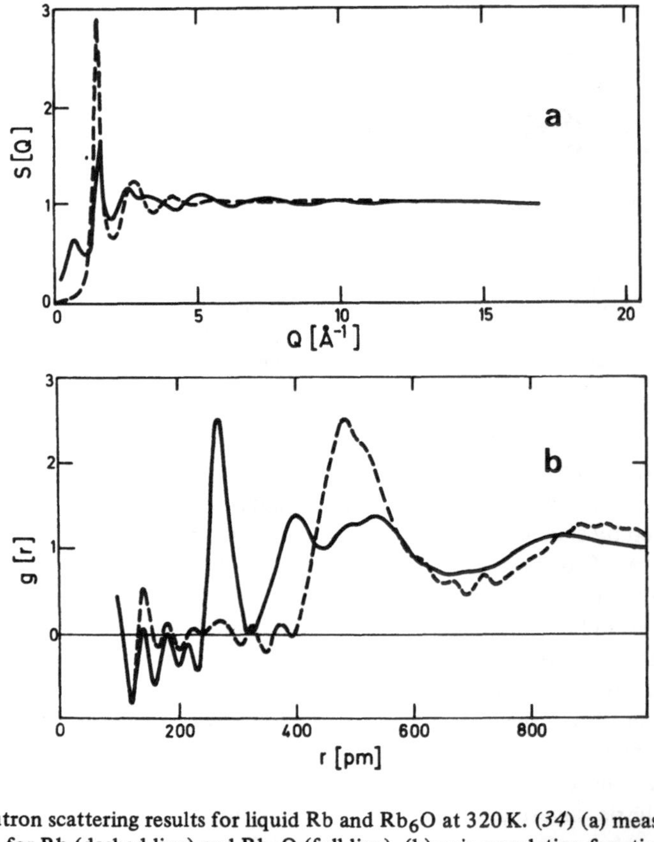

Fig. 13. Neutron scattering results for liquid Rb and Rb_6O at 320 K. (34) (a) measured structure factors S (Q) for Rb (dashed line) and Rb_6O (full line). (b) pair correlation functions g(r) for Rb (dashed line) and Rb_6O (full line)

correllation function derived from S(Q). All peaks at low values of r correspond to interatomic distances, which occur in the solid alkali metal suboxides: The sharp peak at 275 pm is attributed to Rb—O distances, the structure at 400 pm is due to the nearest-neighbor Rb—Rb distances in the clusters, and the structure around 540 pm is significant for intra-cluster Rb—Rb distances along the diagonal of an Rb octahedron, but also stands for inter-cluster distances together with the structures at 480 pm.

These results are in accordance with the assumption of Rb—O clusters in the melt, but the occurrence of the special Rb_9O_2 cluster type is not really proved, as the occurrence of e.g. simple Rb_6O clusters in the melt can not be excluded. Although further neutron and X-ray experiments with variable sample compositions must be obtained, there is indeed some evidence for a decomposition of the Rb_9O_2 cluster in the melt from an investigation of the metastable phase $Rb_{6.33}O$.

Rapid quenching of a melt of composition Rb_7O with liquid nitrogen followed by warming to 140 K (see II.1.) leads to the formation of crystalline $Rb_{6.33}O$ (*17*) (in addition to Rb). The cubic structure of this phase is shown in Fig. 14A, B. The Rb atoms form centered icosahedra, which are arranged in a NaCl-type structure. The Rb—Rb distances within and between the icosahedra are in the range of metallic bonds (483 to 519 pm). Additional Rb atoms, together with the O atoms, occupy positions between the icosahedra. Although the correctness of the assigned O positions

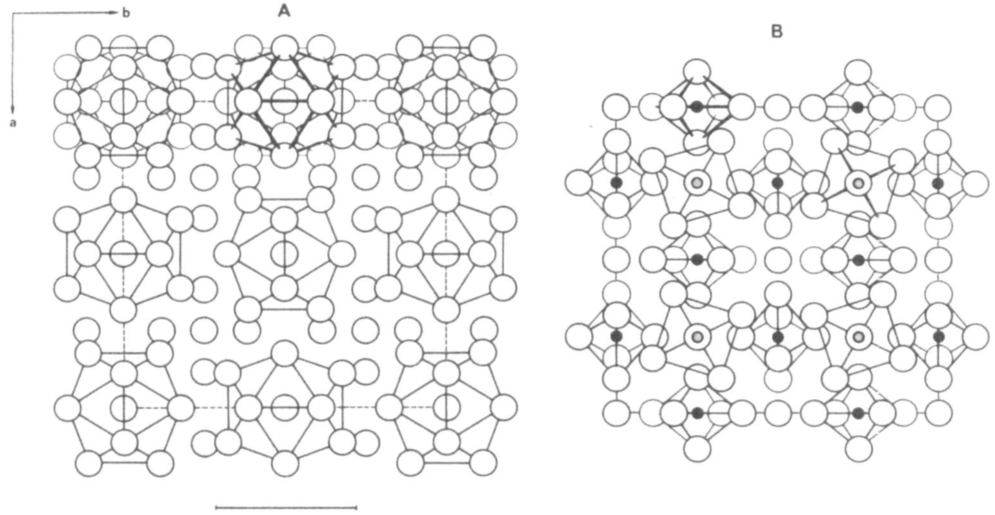

Fig. 14. Projection of the structure of $Rb_{6.33}O$ along [001]. (*17*) Only a layer of a depth c/2 is drawn. (A) Arrangement of centered Rb icosahedra; the icosahedra of the following plane are rotated by 90° as indicated on top of the drawing. (B) Arrangement of Rb_6O coordination polyhedra; (open circles Rb; black and shaded circles O)

still must be verified by neutron diffraction – indeed, the very similar structure of YB_{66} (*35*) offers the same crystallographic problem – the existence of Rb_9O_2 clusters in $Rb_{6.33}O$ can definitely be excluded.

Most probably the O atoms are surrounded by distorted octahedra of Rb atoms [Rb–O distances 270 (2 x) and 360 pm (4 x)], which correspond to the (contracted) quasi-octahedral sites in the bcc lattice of Rb (Fig. 14B). Such isolated Rb_6O units and additional Rb atoms are arranged in a Cu_3Au type lattice according to the formula $(Rb_6O)_3Rb$. The icosahedra, centered by the additional Rb atoms, are exclusively formed from Rb atoms belonging to the Rb_6O units. The occurrence of single Rb_6O units in the metastable intermediate phase $Rb_{6.33}O$ provides evidence that Rb_9O_2 clusters are not the main constituents in the corresponding melt of composition Rb_7O.

IV. Chemical Bonding in the Stable Alkali Metal Suboxides

1. Model of Chemical Bonding

A first insight into chemical bonding within the alkali metal suboxides is gained by comparing the interatomic distances in the compounds Rb_9O_2 and $Cs_{11}O_3$ with those in the "normal" oxides and in the metals Rb and Cs (Fig. 15). The M–O distances nearly correspond to the sum of ionic radii; the pronounced gap between the intra- and inter-cluster M–M distances in Rb_9O_2 and $Cs_{11}O_3$ leads to a bonding model with ionic clusters linked by metallic bonding forces. Such a model follows from simple valence considerations: Assuming the usual valencies (M^+ and O^{2-}), the clusters are charged + 5, which has to be compensated by five (free) electrons in the electroneutral compounds. These additional electrons are definitely essential for the bonding between the clusters as well as for the cluster stability itself.

The compounds Rb_9O_2 and $Cs_{11}O_3$ can be discussed as "complex" metals, a term which means that the simple ions in a "normal" metal are substituted by the complex ionic clusters A and A'. The formation of intermetallic phases of the "complex" metals with additional Rb and Cs leads to the compounds AB_x and $A'B_x$ with purely metallic bonds to the B atoms.

The intermediate position between ionic and metallic type of bonding is clearly reflected in the molar volumes of alkali metal suboxides as it has been discussed in more detail in III.2. The molar volumes of the Rb_9O_2 and $Cs_{11}O_3$ clusters roughly agree with the sums of the molar volumes of M and M_2O, weighted according to stoichiometry; the calculated sums are about 8.5% smaller. (*15*) Both a non-linear decrease in the volume increment of M as a function of charge as well as an increased

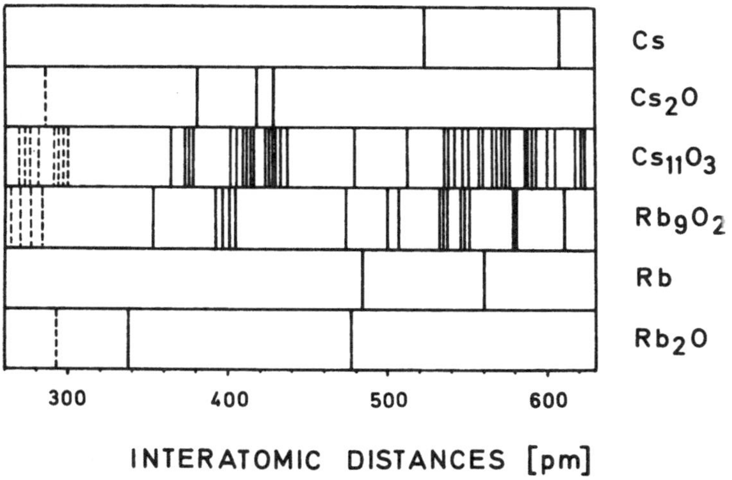

Fig. 15. Distribution of interatomic distances in Rb_9O_2 and $Cs_{11}O_3$ (*15, 24*) compared with the distances in the elements Rb, Cs (*83, 84*) and their oxides Rb_2O (*85*) and Cs_2O. (*86*) Dashed bars correspond to M–O and full bars to M–M-distances

volume of the clusters due to the strong electrostatic repulsion of the O^{2-} ions could explain this deviation. The volumes of the more metal-rich suboxides can be accurately calculated as the sum of the cluster volume and the atomic volume of the B-type atoms taken from the elements Rb and Cs (III.2.).

2. *Structures and Stoichiometries*

It is tempting to look at common features of the structures of alkali metal suboxides to understand the peculiar stoichiometries. One might ask whether certain electronic requirements or special conditions for the packing of clusters and B-type metal atoms force the compositions of the stable compounds.

First of all, the difference between Rb and Cs which form the Rb_9O_2 and $Cs_{11}O_3$ clusters respectively, must be explained. A very qualitative argument is based on the different sizes of the Rb and Cs atoms: As the distribution of M–O distances shows, the O^{2-} ions in the Rb_9O_2 cluster repel each other strongly. (Due to the larger effective charge of O^{2-} in the clusters, the shift of these atoms out of the coordination centers is even more pronounced than with the metal atoms in e.g. $Cr_2Cl_9^{3-}$. (*10*) Obviously, the addition of one more interfering O^{2-} ion is permitted in the case of the larger cage of Cs atoms only. Such size effects might also be the explanation for the non-existence of binary potassium suboxides.

In the description of the different suboxide structures the cluster environments have been given the most attention. This special structural aspect is once more stressed by a compilation of cluster environments in Fig. 16 taken from the rather different but representative structures of Cs_7O, $Cs_{11}O_3$, and $Cs_{11}O_3Rb$.

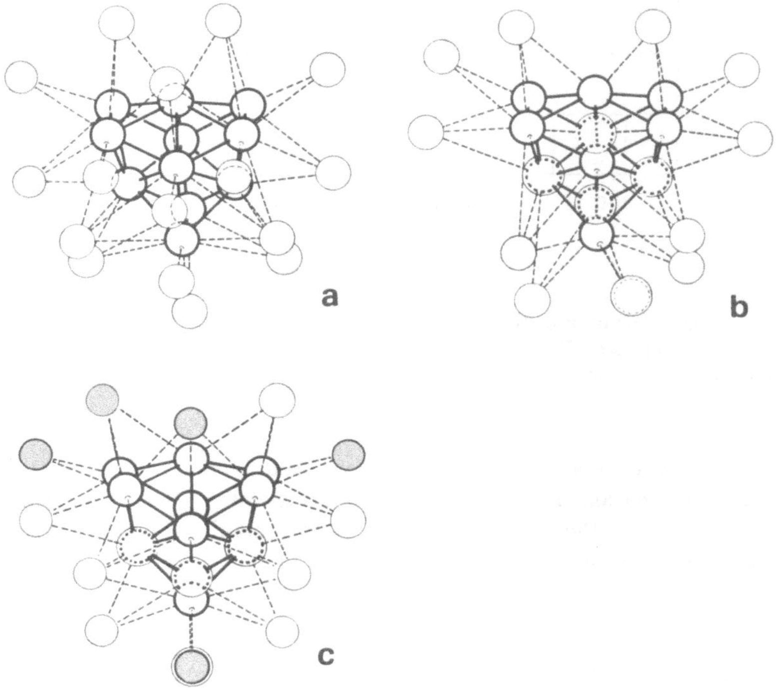

Fig. 16. Atom environments around the $Cs_{11}O_3$ cluster in (a) Cs_7O, *(26)* (b) $Cs_{11}O_3$ *(24)* and (c) $Cs_{11}O_3Rb_7$. *(22)* The Cs atoms belonging to the cluster are stressed by strong lines. The O atoms are omitted from the drawing; Rb atoms are shaded

The metal atoms below the cluster are omitted for the sake of visual clarity. It is obvious that the arrangement of atoms around the clusters in all compounds is strikingly similar. The positions explicitly mentioned with Cs_7O in Section 3.1 are all occupied, although the (ideal) arrangement is rather distorted, when the coordinating atoms belong to adjacent clusters. The irregularity of the environment seems to be most pronounced with the structure of Cs_4O (not shown in Fig. 16). One might even argue that the extreme difficulties in preparing this compound *(11)* could be caused by the bad fitting of the clusters in the structure of Cs_4O.

The characteristic environment of the $Cs_{11}O_3$ cluster in the suboxides reflects the tendency towards a most effective space filling in the structures as it is expected for the metallic bonding between the clusters and the B-type atoms. In fact one may draw the conclusion that space filling is of critical importance for the stoichiometries found with alkali metal suboxides. Obviously only those stoichiometries AB_n are realized, where structures with optimal space filling, based on the characteristic environment of the $Cs_{11}O_3$ clusters, are possible.

3. Ionic Bonding in the Clusters

The bond model for alkali metal suboxides, derived from the crystal structures, finds further support from calculations of electrostatic energies. The Coulombic contribution to chemical bonding, expressed in the Madelung part of lattice energy *(Maple)* *(36–38)* has been calculated for the compounds Cs_7O, Cs_4O and $Cs_{11}O_3$. *(39)* Starting with a charge distribution according to $(Cs^{+2/7})_7O^{2-}$, $(Cs^{+1/2})_4O^{2-}$ and $(Cs^{+6/11})_{11}(O^{2-})_3$, such calculations lead to very small partial contributions of the B-type atoms in Cs_4O and Cs_7O as one would guess. *Maple*-values for these atoms are an order of magnitude smaller than for the Cs atoms in the clusters; B-type atoms carry effectively zero charge. The main contribution to lattice energy comes from the bonding in the $Cs_{11}O_3$ clusters. The *Maple*-values of about 1 140 kcal/mol for the clusters in Cs_7O, Cs_4O and $Cs_{11}O_3$ are convincingly similar such as to show the electrostatic "autarky" of these units. Of course, the chemically nonequivalent positions of the three different kinds of Cs atoms with respect to their coordination by oxygen are reflected in the different *Maple*-values for the three sets of atoms, namely approximately 30 kcal/mol for Cs1, 45 for Cs2 and 60 for Cs3. It is interesting to note that the relative amount of replacement of Cs atoms by Rb described in Section 3.2 is closely correlated to these values.

The ionic bonds inside the Rb_9O_2 and $Cs_{11}O_3$ clusters should be much stronger than the metallic bonds to the cluster environment. This might be concluded from the striking geometrical similarity of the clusters in different compounds, but this argument is not too strong, because the clusters are also in rather similar environments. A direct evidence of the very different strength of intra- and inter-cluster bonds is given by the dynamical behavior of the clusters.

As shown in Fig. 17, the thermal motion of the Cs atoms in the $Cs_{11}O_3$ cluster is markedly anisotropic. The orientations of the thermal ellipsoids with respect to the cluster center are almost the same for all $Cs_{11}O_3$ clusters in the different suboxides. A detailed analysis shows that the thermal vibrations of the Cs atoms occur mainly relative to the O atoms (in addition to a rigid motion of the cluster): The vibration amplitude is largest for the Cs1 atoms and the direction of the largest displacement for these atoms is perpendicular to the strong (but not directional) Cs–O bond. The thermal motion for the Cs2 and Cs3 atoms bound to two and three oxygen atoms,

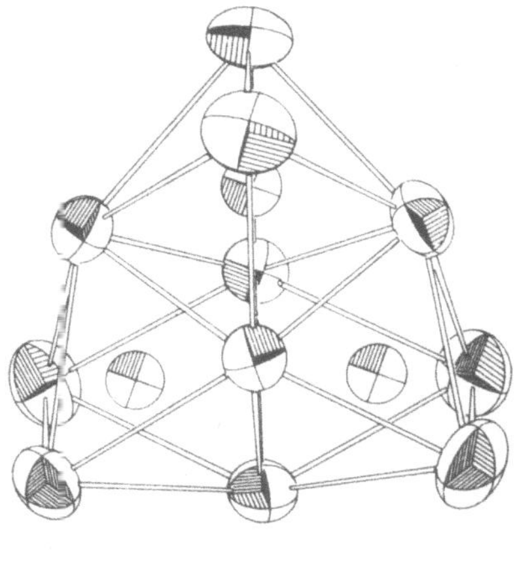

Fig. 17. Thermal ellipsoids of the Cs atoms in $Cs_{11}O_3$ at 300 K. The ellipsoids represent 50% of the electron density *(24)*

respectively, is significantly smaller. The weak bonding of the clusters to their environment of adjacent clusters and B-type atoms is evident from these thermal motions.

Raman spectroscopy with alkali metal suboxides leads to the same conclusion. The spectra of some compounds *(40, 41)* are shown in Fig. 18. Unfortunately, the Raman scattering from the metallic compounds is very poor, especially concerning the weak bands. But obiously the spectra are very similar for all compounds. This is especially true for compounds which contain the $Cs_{11}O_3$ cluster. So it is concluded that the characteristic bands in the spectra correspond to internal modes of the weakly perturbed "molecular" Rb_9O_2 and $Cs_{11}O_3$ clusters. Having in mind the rather weak influence of the matrix on the clusters, it is tempting to describe the clusters as quasi-free ionic aggregates. Two principal questions then arise:

(a) What is the stable geometrical configuration for 9 (11) univalent positive and 2(3) divalent negative ions?

(b) Are the dynamics of such "free" clusters in agreement with the measured data?

Stable configurations of ionic clusters $(M^+X^-)_n$ and $(M^{2+}X^-{}_2)_n$ have been calculated for a number of species on the basis of a simple *Born-Mayer* approach by minimizing the total energy. *(42–44)* The vibrational frequencies evaluated for small clusters of alkali halides compare well with the FIR data of matrix isolated species. *(45)*

Assuming a simple two body interaction

$$V_{ij} = \frac{Z_i Z_j}{r_{ij}} + A \cdot \exp(-r_{ij}/\rho) \quad (A = 34 \cdot 10^{-10}\ \text{erg},\ \rho = 31\ \text{pm})$$

the calculation of stable configurations for the Rb_9O_2 cluster leads to a very interesting result. *(46)* It turns out that an aggregate of 9 Rb^+ and 2 O^{2-} ions according to $[Rb_9O_2]^{5+}$ does not yield a stable configuration, but it loses one cation. Reducing

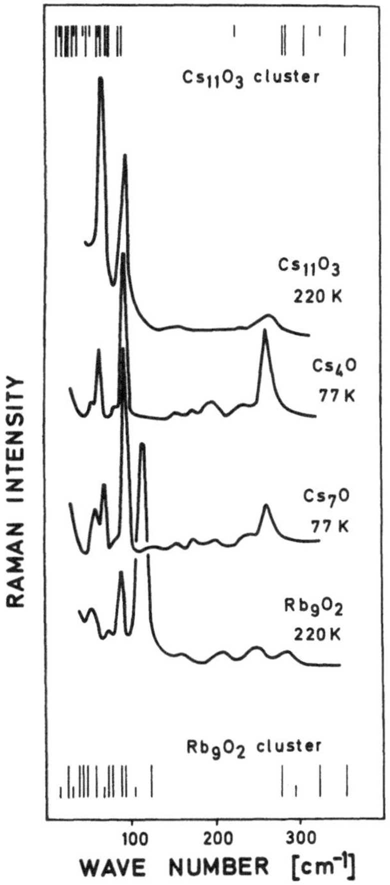

Fig. 18. Raman spectra of alkali metal suboxides (*40, 41*) measured with a 50 mW argon laser. Above the spectra the plot of calculated (*46*) vibrational states for the $Cs_{11}O_3$ cluster is shown, below the same result is given for the Rb_9O_2 cluster. Long bars indicate Raman

the charges of the Rb^+ ions the well known cluster with D_{3h} symmetry is found as the stable configuration along with a metastable cluster with higher energy and D_{4d} symmetry.

Figure 19 illustrates the dependence of the total energy and the interatomic distances on the net charge of the cluster. The energy minimum is reached for the D_{3h} cluster at a net charge $z = +2.5$. It seems important to note that the configuration of lowest energy simultaneously represents the best fit to the observed geometry of the cluster (Table 3).

Similar calculations have been performed for the $Cs_{11}O_3$ cluster with $A = 34 \cdot 10^{-10}$ erg, $\rho = 33$ pm. The calculated stable configuration also corresponds to the observed cluster of D_{3h} symmetry. Although there is a very good agreement between observed and calculated Cs–O distances, rather large deviations occur with the Cs–Cs distances involving the central Cs3. The best agreement is found with a reduced charge of $+0.7$ on these atoms in a cluster of net charge $z = +3.5$.

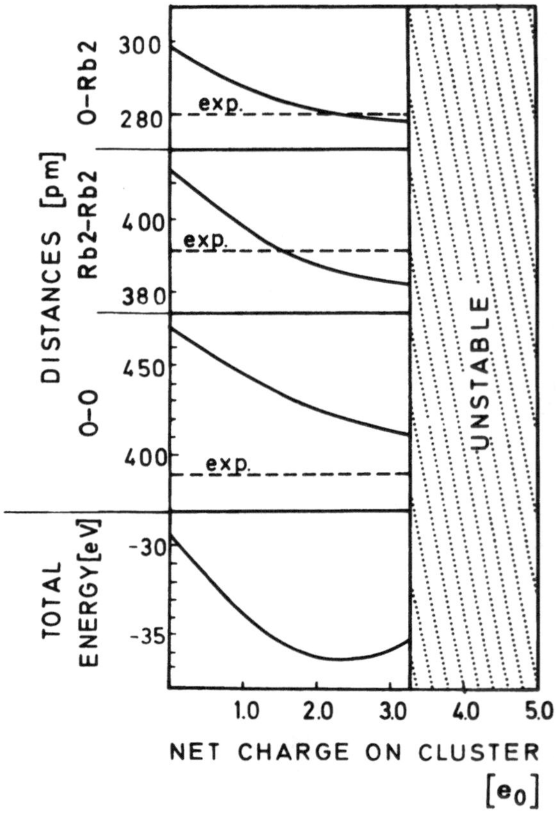

Fig. 19. Calculated total energy and interatomic distances for the Rb_9O_2 cluster as a function of the net charge on the cluster. (*46*) The observed distances are indicated by dashed horizontal lines

Table 3. Comparison of measured interatomic distances [pm] in the clusters Rb_9O_2 and $Cs_{11}O_3$ with distances in ionic clusters $[Rb_9O_2]^{2.5+}$ and $[Cs_{11}O_3]^{3.5+}$, which correspond to configurations of minimized energy

	Rb_9O_2		$Cs_{11}O_3$	
	$d_{Calc.}$	$d_{Meas.}$	$d_{Calc.}$	$d_{Meas.}$
M1–M1	384	394	404	402
M1–M3	–	–	450	414
M2–M2	325	354	–	–
M3–M3	–	–	352	371
M1–O	262	270	273	273
M2–O	279	280	284	289

The validity of the simple ionic bond model is supported by the fact that not only the observed static structures of the clusters are in rather good agreement with the computed stable configurations, but that the dynamic behavior of the clusters is also described by the model. (*46*)

The Rb_9O_2 cluster of ideal symmetry D_{3h} has 13 Raman active ($4\,A_1' + 4\,E'' + 5\,E'$) and 8 IR active modes ($3\,A_2'' + 5\,E'$); the ideal $Cs_{11}O_3$ cluster is characterized by 17 Raman active ($5\,A_1' + 7\,E' + 5\,E''$) and 11 IR active modes ($4\,A_2'' + 7\,E'$). In Fig. 18 the vibrational modes of the clusters calculated on the basis of the simple ionic model for D_{3h} symmetry are plotted. The main features of the spectra are well represented by the calculated densities of vibrational states. The separation between the vibrations of the metal atoms of the clusters at wave numbers below 120 cm^{-1} and the oxygen vibrations around 250–300 cm^{-1} are especially clearly marked out. The intense Raman bands around 100 cm^{-1} most probably are the totally symmetric A_1' modes.

These results, still in a preliminary state, beautifully conform to the intuitive model of chemical bonding with alkali metal suboxides. The description of the characteristic clusters as quasi-free ionic units seems adequate. But in reality the *(in summa)* repulsive forces between the ions have to be compensated by a constraining field of free electrons to make the clusters stable. As a first approximation the constraining field is represented by a shielding effect of the electrons on the Rb^+ ions leading to a reduction of the net charge of the clusters in the described calculations.

Such a modified picture is definitely not contradictory to the formulations $[Rb_9O_2]^{5+}\,5\,e^-$ and $[Cs_{11}O_3]^{5+}\,5\,e^-$ used in Chapter IV.1., because the reduction of the net charges of the clusters does not mean a reduction of the free electron concentration. Indeed, this result is equivalent to an assumption used by Burt and Heine (*47*) to explain the low work functions of $Cs_{11}O_3$ (see V.4.). These authors emphasize that the inner part of the clusters due to the concentrated negative charges of the O^{2-} ions are highly repulsive for electrons. Therefore, the free electrons are confined to the periphery of the clusters. This assumption is visualized by the very simple picture of an ionic character of the inner part of each Cs atom in the cluster and a metallic character of the outer part of each Cs atom.

V. Electronic Properties of Alkali Metal Suboxides

According to the bond model all alkali metal suboxides are metallic. So knowledge on their electrical and electronic properties is essential to verify the model. Electrical resistivity measurements, investigations of optical reflectivities and the UV-photoelectron spectra of alkali metal suboxides are discussed in terms of the bond model in the following section.

1. *Electrical Resistivity of the Stable Compounds*

A number of phases have been measured (*18*) using a probeless technique (*48*). The results are plotted and compared with the pure elements Rb and Cs in Fig. 20. The earlier data of *Brauer* (*49*) differ by less than 10% of the absolute values. The metallic nature of all samples (at least qualitatively) verifies the bond model. The specific resistivity ρ of an alkali metal suboxide is in the same order of magnitude as with the pure elements Rb and Cs (*50*). The compound Cs_7O exhibits a residual resistivity ratio (rrr) defined as ρ (260 K)/ρ (T → O K) in the order of 200, near to the value found for Cs (*50*). The value for the compound $Cs_{11}O_3$ is approximately 50. The temperature dependence of the ideal resistivity (corrected for the residual resistivity) of e.g. Cs_7O is very similar to that of Rb and Cs.

A quantitative discussion of the increase in the resistivities of Rb and Cs upon oxidation reveals some interesting aspects. The resistivity increases by 60% from Rb

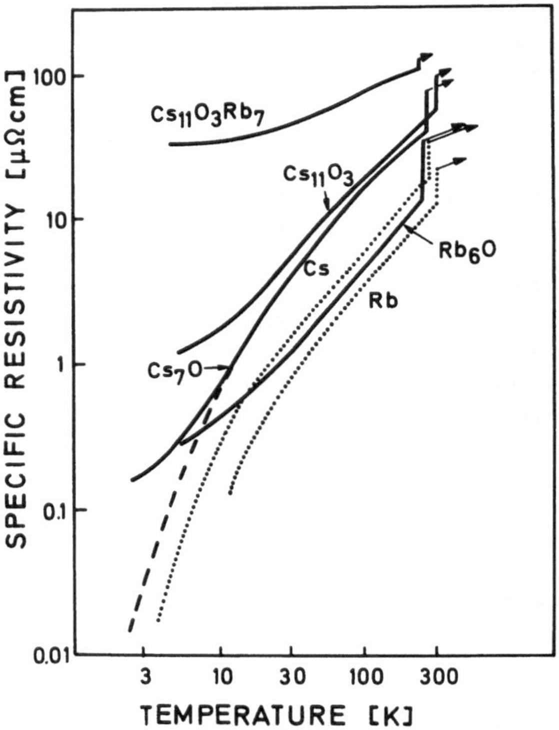

Fig. 20. Electrical resistivities of some alkali metal suboxides compared with the values for the elements Rb and Cs (dotted lines). (*18*) The dashed line with Cs_7O indicates the ideal behavior after subtraction of the residual resistivity. The curve for Rb_6O refers to a sample of this composition only

to Rb_6O and by 150% from Cs to Cs_7O. According to the bond model the free carrier concentration decreases in the same direction. But this decrease is partly compensated by the pronounced contraction of the metal upon oxidation. Thus, the calculated free carrier concentration N decreases by only about 20% when going from Rb (Cs) to Rb_6O (Cs_7O). This small decrease does not explain the increase of ρ. Of course one expects additional scattering of the electrons, e.g. due to the lower symmetry and larger unit cells in the suboxides, but the remarkable difference between the increase of ρ for Rb and Cs, respectively, suggests an additional scattering mechanism, which is more important with Cs suboxides: Cesium behaves as a d-metal, when it is compressed to 30% of its normal pressure volume. *(51)* In fact, the superconductivity of Cs under pressure *(52)* is a consequence of the d-character. It is interesting to note that the very short intra-cluster distances between Cs atoms correspond to those in Cs under 60 kbar pressure (compressed to 35% of the value at normal pressure). Thus s–d scattering might be of significant importance in alkali metal suboxides, but especially in the Cs compounds. Measurements on Cs_7O down to 70 mK, however, give no indication of a superconducting transition; *(53)* the same holds for "Rb_3O" samples down to 300 mK. *(54)*

The electrical properties of the ternary compound $Cs_{11}O_3Rb_7$ are by no means understood completely. The phase has a resistivity, which is an order of magnitude larger than with the other compounds, and a very low value of rrr ~ 3. This behavior is explained as a consequence of the inherent disorder due to the range of homogeneity in this phase (see III.2.). But as it is discussed in V.4. the experimental value of the free carrier concentration in $Cs_{11}O_3Rb_7$ is also significantly smaller than expected.

Measurements of heterogeneous samples have shown that grain boundaries are of little influence as long as suboxides coexist with alkali metal, but become important with a high oxygen content of the samples. Therefore, the published values for "Rb_3O" and "Cs_3O" ($\rho = 21.5\ \mu\Omega$ cm at 223 K *(54)* and 72.1 $\mu\Omega$ cm, respectively) might suffer considerably from grain boundary effects.

The magnetic properties measured for Cs_7O *(26)*, $Cs_{11}O_3$ *(24)* and "Cs_3O" *(29)* are in agreement with the metallic nature of all samples. These suboxides exhibit a small and practically temperature-independent paramagnetism in the order of χ_g = $0.10\ (\pm\ 0.05) \cdot 10^6\ cm^3 \cdot g^{-1}$.

2. *Electrical Resistivity of Metastable Phases*

As mentioned briefly in Section I.1., quenching of melts of the oxidized alkali metals Rb and Cs leads to amorphous samples. The quenching rate of approximately $10^2\ K \cdot sec^{-1}$ to achieve a glass-like state is very low compared to the normal rates of $\sim 10^5\ K \cdot sec^{-1}$ necessary to obtain glassy metals. With certain sample compositions it is possible to obtain amorphous samples, which are free of any crystalline contamination. This holds for Cs_xO ($4 \leqslant x \leqslant 6$); a sample of Rb_7O always contains traces of crystalline Rb. The glass-like and metallic nature of the quenched samples is evident from X-ray, conductivity and PES measurements.

The temperature dependence of ρ is shown for two representative samples (*18*) in comparison to the behaviour of a "classical" glassy metal (*55*) in Fig. 21. The resistivity of the glass-like samples is 5–10 times higher than ρ of the crystalline sample and corresponds fairly well to the extrapolated values of the melt. As long as the low temperature prohibits ordering processes ρ (glass) increases linearly. With many samples a sudden increase occurs, although according to X-ray results the sample is still amorphous. (*18*) This might be due to a transitional state, before a new short-range

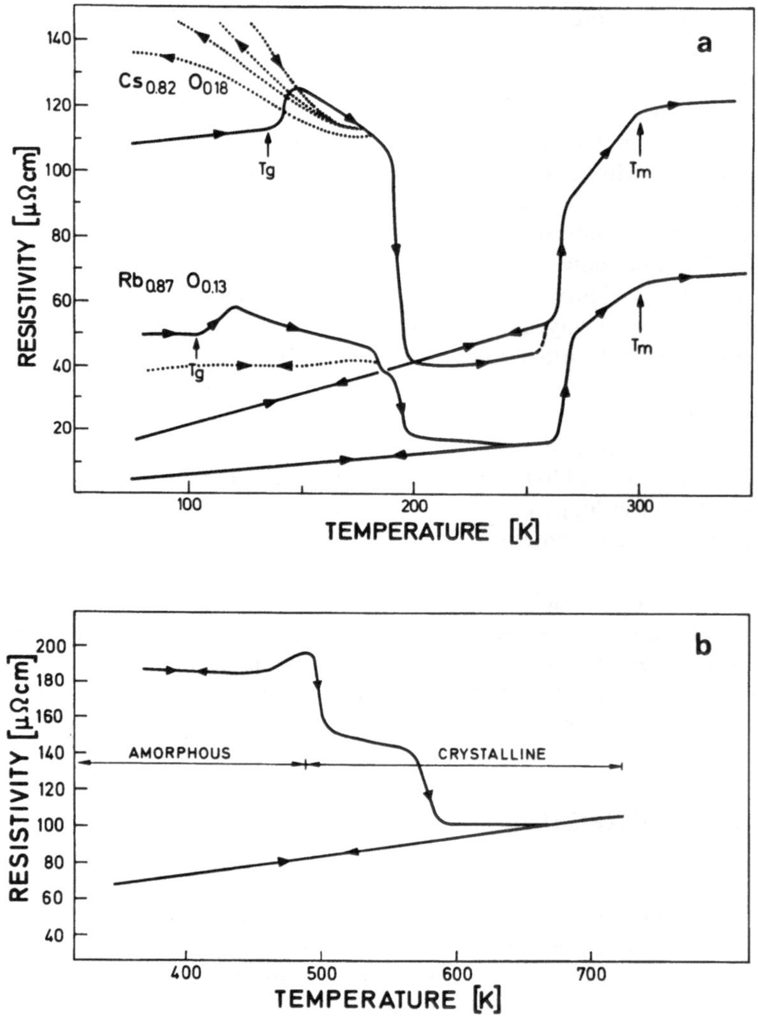

Fig. 21. Electrical resistivities of (a) two quenched alkali metal suboxides, (*18, 20*), which are representative for other compositions, too. For comparison the temperature behavior of (b) an arbitrarily chosen metallic glass is shown (*55*)

order differing from the quenched order in the melt is reached, or due to microcrystal formation with high strain and many phase boundaries. The increase of ρ is followed by a region of a negative temperature coefficient for ρ, before a rapid decrease (often occurring in two steps) indicates the formation of the stable crystalline phases, which sometimes behave strangely as they initially have a smaller resistivity than after an annealing period.

A very remarkable difference is found between the Rb and Cs suboxides chosen in Fig. 21. According to X-ray results $Cs_{0.82}O_{0.18}$ is amorphous below 180 K, whereas $Rb_{0.875}O_{0.125}$ forms the metastable crystalline phase $Rb_{6.33}O$ at about 150 K, which decomposes into Rb + Rb_9O_2 at about 180 K. Now, the formation of $Rb_{6.33}O$ does not appear in any change of ρ, but yet there is a marked difference in the behavior of the metastable amorphous and the metastable crystalline phases indicated in Fig. 21: The temperature coefficient of ρ is negative for the intermediate Cs suboxide (possibly due to strain in the sample). A decrease of temperature results in a resistivity value, which is nearly twice as high as the value for the original quenched sample. The instability of the system is expressed in a gradual decrease of ρ, when the temperature is cycled. In contrast to this behavior, ρ in the quasi-stable system of the Rb suboxide has a weakly positive temperature coefficient and shows no significant time dependence.

Metallic glasses have found much recent interest. Rules have been worked out which should be obeyed by a metallic system to form a glass when the melt is quenched.

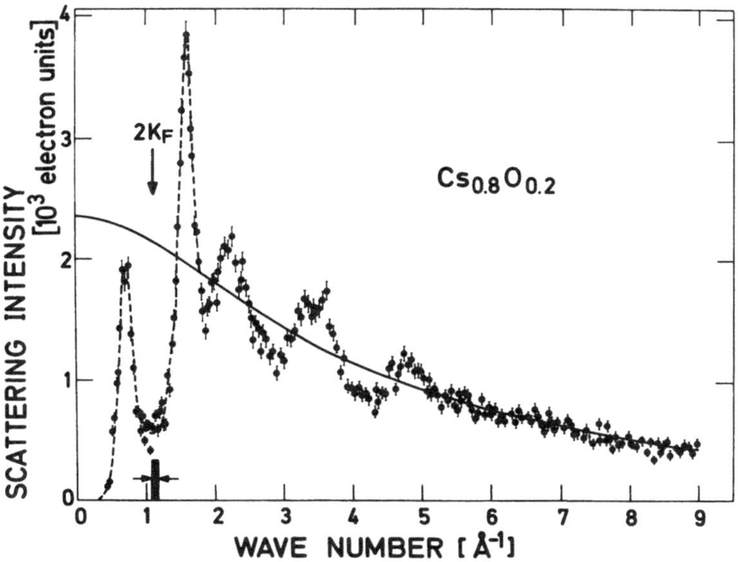

Fig. 22. Scattered X-ray intensity of a glassy sample $Cs_{0.8}O_{0.2}$ at 110 K, normalized to the mean squared scattering factor (solid line). (*20*) The arrow marks $2\,K_F$ indicating the deviation from the condition $2\,K_F = K_p$

These rules essentially result from structural requirements (chemical bond effects (*56*), concept of sphere packings (*57*)) on one side and contributions of the conduction electrons to glass stability on the other side (*58, 59*). Special emphasis is given to the fact that enhanced stability of a metallic glass is always achieved at an average concentration of conduction electrons around 1.7 (± 0.3) per atom. This experimental evidence is explained on the basis of a free-electron model by considering the energy gain for the special condition that the diameter of the Fermi sphere just equals the first large maximum of the structure factor ($2\,K_F = K_p$). (*58*) Although recent results (*60, 61*) provide strong evidence for a dominating role of atomic structure over such electronic effect, the metallic glasses of oxidized alkali metals provide the first examples of the glassy state being stabilized by evenly reducing the conduction electron concentration below 1 electron per atom. As Fig. 22 shows, the condition $2\,K_F = K_p$ is not met, (*20*) but the value of $2\,K_F$ occurs at a minimum of the scattering function for all sample compositions which can be quenched to the amorphous state. Obviously, structural requirements such as the formation of the complex Rb_9O_2 and $Cs_{11}O_3$ clusters and the regular arrangement of these clusters in a crystal are responsible for the easy glass formation with oxidized alkali metals.

This explanation for glass formation with oxidized alkali metals lies in kinetics. In contrast, the electronic rule (*58*) accepted for glassy metals is based on thermodynamics. Indeed, the free electrons in a glassy metal might contribute an additional energy gain, yet this energy gain is not clearly related to the height of the energy barrier between glassy and crystalline state, which is the essential "stability" criterion.

3. The Colors of Alkali Metal Suboxides

The alkali metal Rb is silvery, the metal Cs has a golden color. Upon oxidation these colors change in a characteristic way: Rb becomes golden, and a copper-red is reached with the compound Rb_9O_2. The color of Cs gradually darkens to the luster of bronze with Cs_7O. Cs_4O and $Cs_{11}O_3$ have violet colors and look similar to permanganate. A "Cs_3O" sample of composition $CsO_{0.31}$ has a beautiful blueish-green luster on the free surface, but is dark violet when forming a mirror on a glass wall. A higher oxygen content gradually leads to black samples ($CsO_{0.36}$).

It has been argued (*62*) that the golden color of Cs is no intrinsic property of the element, but is caused by oxygen contamination and formation of dark-colored suboxides. Yet it has been shown (*11*) that the trace of oxygen in a sample $CsO_{0.00001}$ can be detected by the characteristic melting behavior, whereas a measurable change in color does not occur before a composition near $CsO_{0.01}$ is reached.

The origin of the characteristic colors of alkali metal suboxides has been investigated by reflectometry in the range 1.4–6.2 eV (900–200 nm). (*63*) Investigations have been confined to those samples which could be obtained as homogeneous mirrors on quartz glass. Measurements with unpolarized light and vertical incidence are shown in Fig. 23.

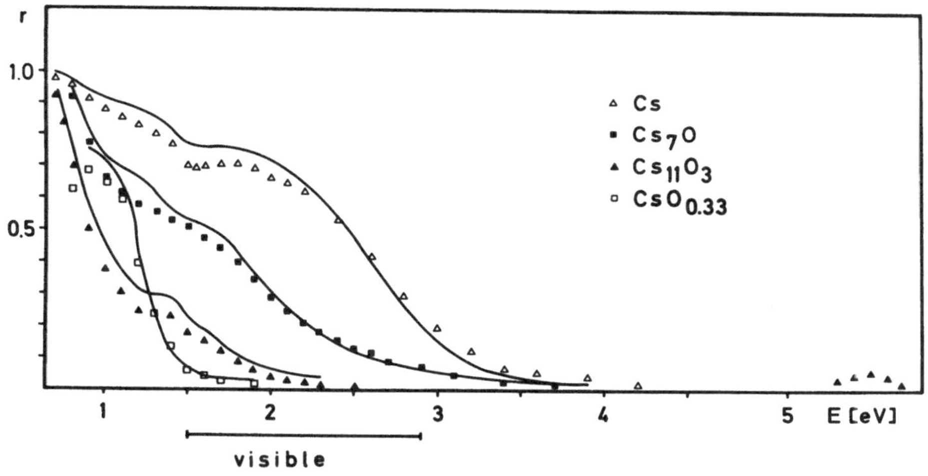

Fig. 23. Reflectivities of mirrors of Cs, Cs_7O, $Cs_{11}O_3$ and $CsO_{0.33}$ on quartz glass at 250 K as a function of the energy of the incident unpolarized light. (*63*) The solid lines are calculated from n and k

Clearly, the plasma edge shifts to lower energy with increasing oxygen content of the samples. The structure of the plasma edge indicates interband transitions; an isolated transition is found with $Cs_{11}O_3$ at 5.5 eV.

A *Kramers-Kronig* analysis of the data (*63*) yields the optical constants n, k. As can be concluded from the reflectivity curves recalculated from n and k, the accuracy is rather low, mainly because of the limitation of data at low energy and beginning interference of the quartz at highest energies. The preliminary character of these reflectivity measurements is evident.

Yet the measurements clearly reveal that the color of Cs as well as the gradual darkening and development of characteristic colors upon oxidation is essentially connected with the frequencies of the plasma oscillations. These cause metallic reflectivity in the whole spectrum of visible light with Cs, Cs_7O and $Cs_{11}O_3$, but reflectivity rapidly decreases especially on the low energy side of the spectrum. In contrast to Cs, Cs_7O and $Cs_{11}O_3$ the phase $CsO_{0.033}$ is transmittant for blue light. Its reflectivity for the rest of the visible spectrum is very low.

Energy absorbing processes other than plasma oscillation influence the structure of the plasma edge of Cs and its suboxides. Indeed, deviations from the "quasifree electrons" postulate are known for cesium. (*64*) Deviations are evident for the suboxides, too, as the dielectric function ϵ_1 (E) exceeds the value $\nu_1 = 1$ at high energy. So it is not surprising that the dielectric loss functions plotted for Cs and its suboxides in Fig. 24 yield maxima with complicated structures. But as plasma oscillations are definitely the main contribution to the energy absorbing processes, the energy loss structures can be used to control the bonding model for alkali metal suboxides on a semi-quantitative scale.

Within the simplified picture of free electrons, the measured plasma frequencies ν_1 taken from the absorption maxima in Fig. 24 have been used to calculate the free carrier concentration N according to $\nu_1^2 = \frac{c^2 N}{\pi \cdot m}$. For an easy comparison, Table 4

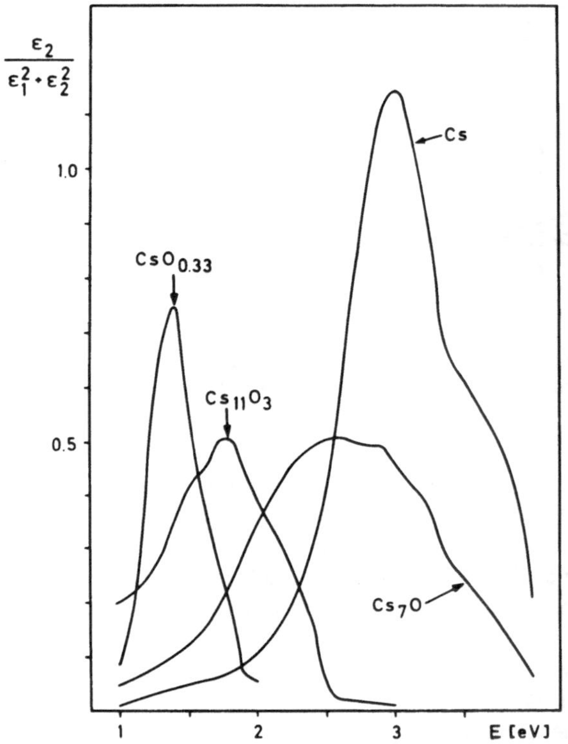

Fig. 24. Dielectric loss functions for Cs, Cs_7O, $Cs_{11}O_3$ and Cs_3O from reflectivity data. *(60)* The absorption maxima yield approximate values for the plasma frequency ν_1

Table 4. Plasma frequencies determined from reflectivity data for Cs and some suboxides. *(63)* In the case of the cluster compounds Cs_7O and $Cs_{11}O_3$ the agreement between experimental (n_0) and expected numbers of free electrons (n_c) is convincing. A large deviation is found for "Cs_3O". The small value of n_0 most probably reflects an enrichment of Cs at the phase boundary between the sample and the glass

	$\nu_1[10^{14}\,sec^{-1}]$	$n_0[e_0]$	$n_c[e_0]$
Cs	7.3	0.8	1
$Cs_{11}O_3 \triangleq Cs_7O$	6.3	10.3	15
$Cs_{11}O_3$	4.5	2.4	5
"Cs_3O"	3.4	0.4	1

shows the values n_0 (electrons per formula unit) which result from the experiment and the values n_c expected according to the bond model. The decrease in the number of free electrons (relative to Cs) roughly agrees with the expected variation. Yet it is clear that further measurements and a more sophisticated evaluation procedure are necessary to use optical data for a quantitative verification of the bond model.

4. *Photoelectron Spectroscopy*

XPS (ESCA) and UPS are powerful tools to determine electronic energy levels by irradiating a sample with monochromatic X-rays and UV light, respectively, and measuring the energy distribution of the emitted electrons. With alkali metal suboxides UPS provides a quantitative proof of the bond model. Furthermore, the UPS results offer an explanation for the low energy photoemission process with oxidized Cs and thus (unexpectedly) open up a field of applied research with infrared sensitive photocathodes. The discussion of the results will be focussed on the chemical bonding.

XPS is usually preferred – at least by chemists – with solid samples because of the very small escape depth of electrons with UPS and the resulting danger of false

Table 5. Binding energies [eV] of the core levels in Rb and Cs and their suboxides referred to the Fermi level $E_F = 0$ from XPS (Al Kα) and UPS. *(40, 65, 66)* The underlined data from UPS have an accuracy of 0.1 eV. Letters A and B designate cluster and B-type atoms

	Rb	Cs	$Cs_{11}O_3Cs_{10}$	$Cs_{11}O_3$	$Cs_{11}O_3Rb_7$
O 2p	–	–	2.7	2.7	2.7
1s	–	–	531	–	532
Rb $4p_{3/2}$	15.2	–	–	–	15.3
$4p_{1/2}$	16.1	–	–	–	16.2
$3d_{5/2}$	112.0	–	–	–	111.9
$3d_{3/2}$	113.0	–	–	–	113.0
$3p_{3/2}$	239.1	–	–	–	240.2
$3p_{1/2}$	248.7	–	–	–	248.9
3s	326.7	–	–	–	–
Cs $5p_{3/2}$	–	12.1	11.5 (A); 12.2 (B)	11.6 (A)	11.5 (A)
$5p_{1/2}$	–	14.0	13.2 (A); 14.0 (B)	13.3 (A)	13.1 (A)
$4d_{5/2}$	–	77.5	77.8	–	77.4
$4d_{3/2}$	–	79.8	79.8	–	79.3
$4p_{3/2}$	–	161.3	161.3	–	161.0
$4p_{1/2}$	–	172.4	172.9	–	171.9
4s	–	232.3	–	–	232.2
$5d_{5/2}$	–	726.6	726.6	–	726.2
$3d_{3/2}$	–	740.5	740.7	–	740.1
$3p_{3/2}$	–	1003.0	–	–	–
$3p_{1/2}$	–	1071.0	–	–	–

information due to surface effects. A part of the Al Kα spectra for Cs, Cs_7O, $Cs_{11}O_3Rb_7$ and Cs_2O *(40)* at the energy of the 3d levels of Cs is shown in Fig. 25. The $3d_{5/2}$ and $3d_{3/2}$ peaks are symmetric with Cs_2O and Cs. In the case of the metal the main structures are accompanied by clearly resolved energy loss structures due to the excitation of the first and second plasmon. With the suboxides these structures are not resolved. There is only a pronounced assymmetry of the peaks. The energies of the core levels with respect to the Fermi level are listed in Table 5. Obviously, the low resolution of XPS renders a discussion of energy differences between electronic levels in Cs and the suboxides impossible.

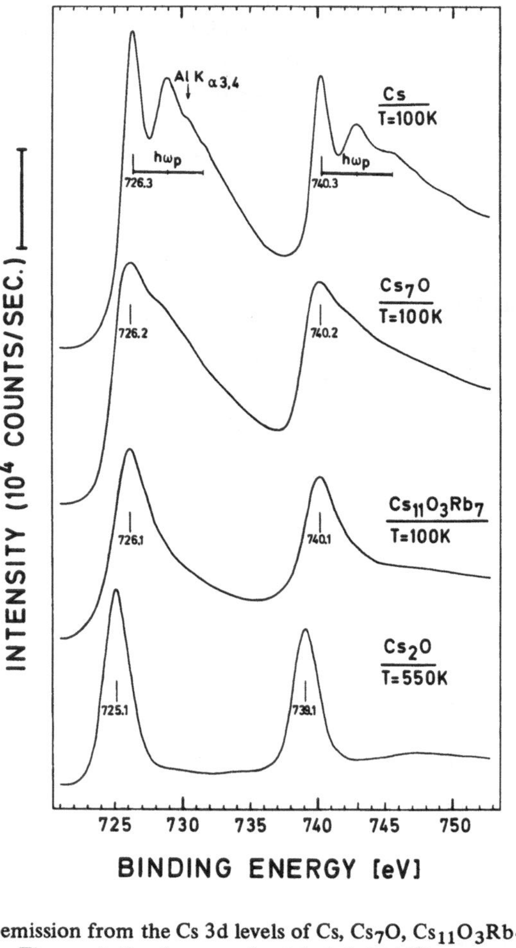

Fig. 25. Photoemission from the Cs 3d levels of Cs, Cs_7O, $Cs_{11}O_3Rb_7$ and Cs_2O, excited by Al Kα radiation. The resolution is approximately 1.2 eV *(40)*

Much more detailed information is gained from UPS data. Experimental details are described elsewhere. *(40, 65, 66)* It should only be noted that the samples have been crystallized in the PE spectrometer at base pressures of approximately 10^{-9} mbar and that the surface of the samples has been scraped prior to and repeatedly in-between measurements. The mechanical cleaning of the sample surface proves to be

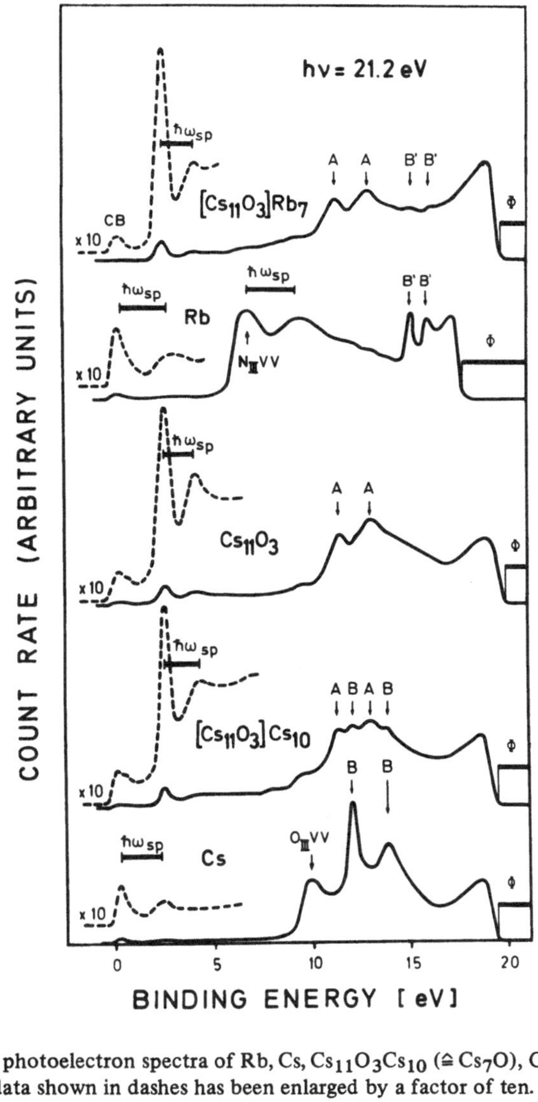

Fig. 26. HeI photoelectron spectra of Rb, Cs, $Cs_{11}O_3Cs_{10}$ (≙ Cs_7O), $Cs_{11}O_3Rb_7$ and $Cs_{11}O_3$ at 100 K. The data shown in dashes has been enlarged by a factor of ten. The resolution is 0.1 eV. The assignments to the structures are discussed in the text

essential to get reproducible results. Spectra have been recorded at 16.8 (NeI), 21.1 (HeI), 26.9 (NeII) and 40.8 eV (HeII). The main features of the photoemission can be described on the basis of the HeI spectra.

The spectra of Rb, Cs, Cs_7O, $Cs_{11}O_3Rb_7$ and $Cs_{11}O_3$ measured with the HeI resonance line are shown in Fig. 26. The spectra exhibit a lot of structure superimposed on a smooth background of secondary electrons, whose intensity increases towards low kinetic energy. All special features of the spectra can be explained and will be discussed in the following order:

(a) conduction band,
(b) oxygen 2 p level,
(c) core levels of the metal atoms,
(d) Auger transitions,
(e) energy loss structures and
(f) changes in work functions.

(a) The partly occupied conduction bands (CB) clearly indicate the metallic character of all samples. The energy scale refers to the Fermi level at 0.0 eV binding energy. The shape of the conduction band in Rb and Cs essentially does not change when the excitation energy (and thereby the escape depth) is varied. In agreement with calculations (*67*) for a half-filled s band the shape of the surface and bulk density of states is the same. The shape of the band indicates the nearly free electron behavior. As Fig. 27 shows, the measured HeI spectra suggest a more complicated shape of the conduction band to occur in alkali metal suboxides at first glance. Although the calculated density of states (*47*) for $Cs_{11}O_3$ looks similar to the experimental results, the additional maximum below the Fermi level is evidently caused by photoemission from the 02 p level excited by the 23.03 eV satellite line of HeI radiation.

Assuming a free electron-like behavior of the valence electrons remaining at the alkali metal atoms in the suboxides, the experimentally determined width of the conduction band provides a measure for the electron density and thus a proof of the bond model.

The Fermi energy E_f, which corresponds to the energy difference between the bottom of the conduction band and the Fermi level, yields the electron density N/V according to (*68*)

$$\frac{N}{V} = \frac{(2\,m \cdot E_f)^{3/2}}{3\,\pi^2 \cdot \hbar^3}\,.$$

The measured Fermi energies of the elements are slightly smaller than the values given by other authors (*69*) and lead to electron concentrations n_0, which are approximately 90% of the expected values. Assuming a similar deviation for the suboxides, the experimental values are in very good relative agreement with the expected values (Table 6). Only in the case of $Cs_{11}O_3Rb_7$ is the concentration of free electrons found to be too low (see V.1.). Of course, the experimental error in the determination of E_f is rather large. A more accurate determination of the electron concentration via the energy loss structures is discussed in (e).

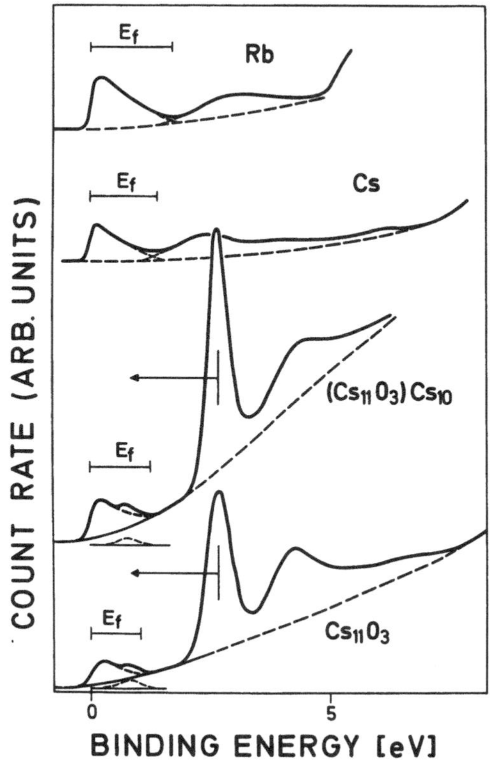

Fig. 27. Fermi energies E_f derived from the HeI spectra. *(40, 66)* In the case of the suboxides the photoemission from the O 2p level due to the 23.09 eV satellite line causes an additional structure in the conduction band region indicated by arrows, which has to be corrected

Table 6. Fermi energies E_f observed for Rb and Cs and their suboxides in the HeI spectra. *(40, 66)* The numbers of free electrons n_0 are calculated from these values and compared with the values n_c, which are expected from the bond model

		For one formula unit		
	E_f[eV]	V[10^{-24} cm^3]	n_0[e_0]	n_c[e_0]
Rb	1.7	89	0.91	1
Cs	1.4	111	0.90	1
$Cs_{11}O_3$	1.05	960	4.8	5
$Cs_{11}O_3Cs_{10}$	1.25	2095	13.5	15
$Cs_{11}O_3Rb_7$	1.2	1600	9.7	12

(b) The alkali metal suboxides are characterized by a very narrow peak (half-width approximately 0.6 eV) at 2.7 eV binding energy due to photoemission from the oxygen 2p levels. The spin orbit splitting of these levels cannot be observed, as it is smaller than the experimental resolution of 0.1 eV.

The binding energy of 2.7 eV for the O 2p electrons is unusually small compared to Na_2O (4.2 eV) (*70*) or transition metal oxides (4–10 eV) (*71*). Indeed, it is the smallest known value which has to be interpreted as due to the comparably weak electrostatic field of positive charges acting on the O^{2-} ion in the suboxides. The very narrow structure of the O 2p band is unusual, too. A band width of 5 eV is reported e.g. for NbO. (*73*) The narrow O 2p band is characteristic of crystalline as well as amorphous suboxides. Obviously, it is the well-defined atomic environment and the isolation of the nearly "gas-like" O^{2-} ions in the quasi-molecular clusters, which is responsible for the narrow O 2p band in the suboxides, as this band is significantly broadened with the oxide Cs_2O itself.

A very narrow O 2p band is observed with earlier measurements (*74*) on partially oxidized Cs films, and the result has been discussed in terms of isolated O^{2-} ions incorporated into the films under a surface layer of metallic cesium. This interpretation is very near to reality. Unfortunately, these investigations have been performed with 10.2 eV excitation energy. So the core levels of Cs are not recorded; but the measurements definitely refer to Cs suboxides. In this respect it is interesting that during the early stages of oxidation of K (*75*) and Sr (*76, 77*) very narrow O 2p bands are observed, although bulk suboxides of these materials are not yet known.

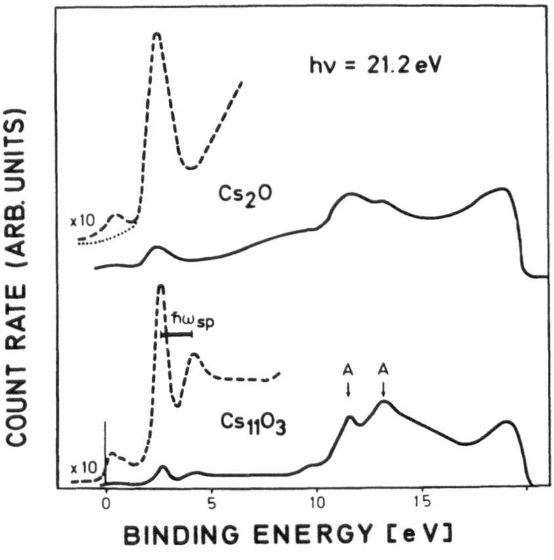

Fig. 28. HeI photoelectron spectra of Cs_2O and $Cs_{11}O_3$.(*40, 66*) For explanations see Fig. 26. The peak at approximately 1 eV in the spectrum of Cs_2O is due to the emission from the O 2p level excited by the 23.09 eV satellite line (see Fig. 26)

(c) The spin-orbit split Rb 4p and Cs 5p) levels (designated B′ and B, respectively, in the spectra of Rb and Cs in Fig. 26) clearly reflect the different chemical bonding of the metal atoms in alkali metal suboxides. In the spectrum of Cs_7O ($\triangleq Cs_{11}O_3Cs_{10}$) the 5p doublet occurs at binding energies as with Cs metal, but another doublet is observed at lower binding energies. One immediately associates the doublet at higher binding energies with the photoemission from the purely metallic B-type atoms in the structure of Cs_7O, whereas the shifted doublet corresponds to the atoms in the clusters A.

This assignment is verified by the spectra of $Cs_{11}O_3$ and $Cs_{11}O_3Rb_7$. Both compounds contain only Cs atoms incorporated in clusters. Therefore, only the Cs 5p doublet at lower binding energies is observed. The negative shift (lowering of the binding energies) for the 5p electrons of the cluster atoms is a somewhat surprising fact. In a free ion approximation one would expect the binding energy of the electrons of a positively charged ion to be larger than that of the neutral atoms. Obviously, it is the field contribution which leads to the decrease in binding energies for the cluster atoms. The field contribution is quite different for the atoms Cs 1, 2 and 3, as expressed in the different values of the partial Madelung energies for these atoms (see IV.3.). Consequently, the Cs 5p bands of the cluster atoms appear as significantly broadened structures in the spectra of $Cs_{11}O_3$ and $Cs_{11}O_3Rb_7$.

Similar negative shifts with respect to the metals have been observed for Ag_2O (*78*), AgCl (*79*) and AgBr (*80*) as well as for Cu oxides (*81*). These results are still a matter of controversy, because all samples are semiconducting and hence the absolute values of the binding energies are difficult to obtain. With the Cs suboxides the negative shifts have been proved relative to the Fermi level as well as to the "internal standard" of the B-type metal atoms.

(d) In the spectra of Fig. 26 *Auger* transitions resulting from core holes in the N and O shell of Rb and Cs, respectively, are observed. The spectrum of Rb is characterized by a very strong $N_{III}VV$ transition, corresponding to holes in the $4p_{3/2}$ band, which are filled by electrons from the conduction band. A faint indication of the $N_{II}VV$ transition due to holes in the $4p_{1/2}$ band is observed. With Cs only the $O_{III}VV$ transition occurs. Obviously the holes in the $5p_{3/2}$ band are filled by electrons from the $5p_{1/2}$ band, giving rise to a life-time broadening (*82*) of the $5p_{1/2}$ band and a decrease in peak height for the $5p_{1/2}$ structure in the spectrum. The same kind of *Coster-Kronig* transition explains the weakness of the $N_{II}VV$ structure in the Rb spectrum.

In the spectra of the suboxides the structures caused by *Auger* electrons are of minor importance. The decrease in intensity most probably is due to the reduced density of states in the conduction band as well as the much broader structures of the core levels.

(e) All investigated samples are metals with the exception of Cs_2O. It is a well-known fact that electrons which pass through a metal lose energy while creating plasmons. The energy of a photoelectron can be reduced by some multiple of the energy of a volume ($\hbar\omega_p$) or surface plasmon ($\hbar\omega_{sp}$), before it is emitted from the surface, leading to energy loss peaks in the PE spectrum. Such energy loss peaks due to plasmon excitations accompany every peak in the spectra at virtually higher binding energies (lower kinetic energies).

It depends upon the energy of the exciting radiation (determining the escape depth of the photoelectrons), whether the energy loss mainly corresponds to $\hbar\omega_p$ or $\hbar\omega_{sp}$. With HeI radiation losses due to surface plasmons are observed, in the given Al $K\alpha$ spectra creation of volume plasmons is the dominant process (see Fig. 25), and with HeII both processes are observed separately. In the HeI spectra of Rb and Cs the energy losses $\hbar\omega_{sp}$ are measured for the conduction band. In the case of the suboxides the energy loss peak of the O 2p structure is particularly suited to determine $\hbar\omega_{sp}$.

In Fig. 29 the energy loss structures at $\hbar\omega_p$ and $\hbar\omega_{sp}$ are shown for Cs_7O, which has been excited with HeII radiation, as well as the loss structures due to $\hbar\omega_{sp}$ and $2\hbar\omega_{sp}$ for a sample of approximate composition Cs_7O, excited by HeI radiation.

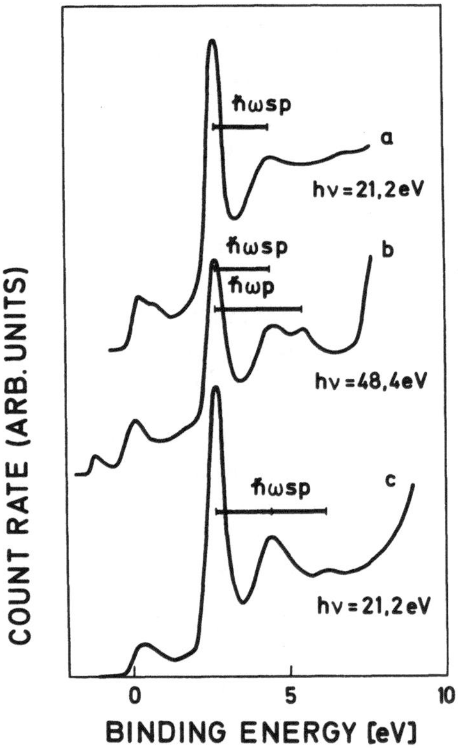

Fig. 29. Energy loss structure related to the O 2p band in Cs_7O (a, b) as observed with HeI (surface plasmon) and HeII radiation (volume and surface plasmon). Curve (c) shows the excitation of two surface plasmons in the HeI spectrum with a sample of approximate composition Cs_7O (*66*)

Any discussion of these structures in terms of shake up processes of adsorbed molecules can thus be rejected.

Assuming an equivalent decrease of the electron concentration to occur in the surface of a suboxide as found for the elements Rb and Cs, the corrected electron concentrations n_0' are derived, which agree very well with the expected values n_c for the binary Cs suboxides, thus verifying the bond model. In the case of the ternary compound $Cs_{11}O_3Rb_7$ the electron concentration n_0' is considerably lower than expected. Further measurements are necessary to understand the interesting process of electron trapping in this phase (see V.1.). and perhaps in other Rb-Cs suboxides.

The values of $\hbar\omega_{p,sp}$ decrease with increasing oxidation state of the alkali metal due to the decrease in the free carrier concentration. According to the equation

$$\omega_{sp}^2 = \frac{1}{2 + \Delta\epsilon_c} \frac{4\pi N e^2}{m^*},$$

neglecting deviations from the free electron mass, but taking into account contributions $\Delta\epsilon_c$ due to core polarizabilities and interband transitions, the numbers N of free electrons are calculated (*65*) from the measured values ω_{sp}. These values are taken from the conduction band structures for Rb and Cs, and from the O 2p structures for the suboxides. The results are summarized in Table 7. The energy of the surface plasmon $\hbar\omega_{sp}$ decreases from 2.0 to 1.55 eV upon oxidation of Cs to $Cs_{11}O_3$.

Table 7. Observed values of the surface plasmon energy $\hbar\omega_{sp}$ for Rb, Cs and some suboxides from HeI spectra. (*40, 65*). The concentration of free electrons N is derived from $\hbar\omega_{sp}$ and yields the experimental number of electrons per formula unit n_0 which is corrected to set n_0' (Rb, Cs) = 1. The expected values n_c correspond to the bond model

			For one formula unit			
	$\hbar\omega_{sp}$ [eV]	N [10^{21} cm^{-3}]	V [10^{-24} cm^3]	n_0 [e_0]	n_0' [e_0]	n_c [e_0]
Cs	2.0	6.9	111	0.8	1.0	1
$Cs_{11}O_3$	1.55	4.3	960	4.1	5.1	5
$Cs_{11}O_3Cs_{10}$	1.75	5.4	2095	11.2	14.0	15
Rb	2.3	8.6	89	0.8	1.0	1
$Cs_{11}O_3Rb_7$	1.55	4.2	1600	6.7	8.4	12

(f) Last, but not least, the spectra shown in Fig. 26 yield one further interesting piece of information about alkali metal suboxides: From the spectra approximate values for the work functions can be extracted.

Photoelectrons with kinetic energies smaller than the work functions cannot escape into vacuum. In the spectra of Fig. 26 the photoelectrons at the highest observable binding energies correspond to E_{kin} = 0 eV and the differences to the full range of excitation energy 21.2 eV are equal to the work functions Φ of the metallic samples, as (fortunately) the work function of the spectrometer is larger than Φ of the suboxides. The approximate values of Φ are 1.95, 1.75 and 1.35 eV for Cs, Cs_7O and $Cs_{11}O_3$. Thus, oxidation of Cs leads to a decrease of the work function in agreement with earlier measurements. (*74*)

The experimental values of $\hbar\omega_{sp}$ lead to the observed electron concentrations n_0 (again referred to one formula unit), when using average values for $\Delta\epsilon_c$ and the information about volumes taken from the crystal structures. (*65*) The expected values n_c are calculated on the basis of the bond model (Table 7). The experimentally derived values n_0 are 20% smaller than the expected values n_c for the pure elements. In spite of the simplifications in the calculations, e.g. taking m instead of m*, the determined electron deficiency seems to be real, as a decreased electron concentration is also calculated from the surface plasmon energies measured for Rb and Cs by *Kunz*. (*87*)

The drop in the work function for cesium suboxides is explained as being caused by a quantum size effect. (*47*) The electronic structure of the suboxides is discussed in terms of a simple structural model with clusters occupied by the O^{2-} ions. The interior of the clusters is highly repulsive for the conduction electrons and the clusters are separated within a Fermi wave length. These model calculations lead to a decrease in work function with respect to Cs by 0.1 and 0.9 eV for Cs_7O and $Cs_{11}O_3$, respectively.

VI. Concluding Remarks

The very low work functions of alkali metal suboxides suggest interesting chemical and physical applications. The reducing power of the suboxides should be even stronger than that of the pure metals Rb and Cs. But no chemical experiment has been performed so far to prove this assumption. From a physical point of view the low work functions make Cs suboxides very interesting photocathode materials.

Indeed, it has been demonstrated (*88*) that Cs suboxides are of essential importance in the famous infrared sensitive S1 photocathode (*89*). These cathodes contain a thin layer of oxidized Cs on a silver substrate. It is evident from UPS measurements on such a cathode prepared in a PE spectrometer that the Cs–O layer is essentially composed of $Cs_{11}O_3$, or a higher oxidized, but still metallic suboxide, giving rise to the characteristic spectrum of the bulk material, although the layer has only the thickness of a few atoms. (*88*) The high yield of photoelectrons in the near infrared with the S1 photocathode results from two effects: $Cs_{11}O_3$ is characterized by a sufficiently small work function (Φ = 1.35 eV) to emit photoelectrons when irradiated with infrared light. Furthermore, the energy necessary to create surface plasmons is also very low ($\hbar\omega_{sp}$ = 1.55 eV). Therefore, photoemission is enhanced due to surface plasmon decay when the energy of the incident light is in the order of 1.55 eV. For further details one is referred to the original work. (*88*)

Applicational aspects of the alkali metal suboxides in photocathodes have not been discussed in great detail, as this work concentrates on another aspect, namely demonstrating the intimate relations between the model for chemical bonding and the resulting structural and physical properties of this unique class of compounds. Initially it was simply the puzzling composition of the compounds which stimulated this investigation, but the variety of unusual properties exhibited by the alkali metal suboxides in the course of the work made them a fascinating object of study.

Acknowledgement. It is a pleasure to thank all those who contributed to the referred work on alkali metal suboxides, especially *W. Brämer, H.-J. Deiseroth, B. Hillenkötter* and *E. Westerbeck* for the preparative and structural work, *T.P. Martin* and *H. J. Stolz* for the work on cluster stability and dynamics, *W. Bauhofer, G. Ebbinghaus, W. Braun* and *K. Mack* for the investigation of electronic and optical properties. The careful drawing of the figures (*D. Skuras* and *R. Burger*) and typing of the manuscript (*I. Claussen* and *J. Rausch*) is gratefully acknowledged. Thanks are also addressed to the "Deutsche Forschungsgemeinschaft" and "Fonds der Chemischen Industrie" who supported this work.

References

1. *Schäfer, H., v. Schnering, H. G.:* Angew. Chem. *76*, 833 (1964)
2. *Cotton, F.A.:* Quart. Rev., Chem. Soc. *20*, 389 (1966)
3. *King, R.B.:* Progr. Inorg. Chem. *15*, 287 (1972)
4. *Vahrenkamp, H.:* Structure and Bonding *32*, 1 (1977)
5. *Corbett, J.D.:* Chemistry. Vol. 21, Edit. Lippard. New York: John Wiley & Sons, Inc. 1976, p. 129
6. *Lokken, D.A., Corbert, J.D.:* Inorg. Chem. *12*, 556 (1973)
7. *Simon, A., Mattausch, Hj., Holzer, N.:* Angew. Chem. *88*, 685 (1976) [Intern. Ed. *15*, 624]
8. *Rengade, E.:* CR. hebd. Séances Acad. Sci. *148*, 1199 (1909)
9. *Rengade, E.:* Bull. Soc. Chim. France [4] *5*, 994 (1909)
10. *Simon, A.:* Homoatomic rings, chains and macromolecules of main-group elements. Rheingold, A. L. (ed.). Amsterdam: Elsevier 1977, p. 117
11. *Simon, A.:* Z. anorg. allg. Chem. *395*, 301 (1973)
12. *Simon, A.:* Crystal structure and chemical bonding in inorganic chemistry. Rooymans, C. J. M., Rabenau, A. (eds.). North Holland Publ. Co. 1975, p. 47
13. *Touzain, Ph.:* Can. J. Chem. *47*, 2639 (1969)
14. *Touzain, Ph.:* Rev. Chim. Minér. *8*, 277 (1971)
15. *Simon, A.:* Z. anorg. allg. Chem. *43*, 5 (1977)
16. *Simon, A., Deiseroth, H.-J.:* Rev. Chim Minér. *13*, 98 (1976)
17. *Deiseroth, H.-J., Simon, A.:* Z. Naturforsch., *33b*, 714 (1978)
18. *Bauhofer, W., Simon, A.:* Z. anorg. allg. Chem., in press
19. *Westerbeck, E.:* Thesis, Münster (1975)
20. *Bauhofer, W., Simon, A.:* Phys. Rev. Lett., *40*, 1730 (1978)
21. *Simon, A., Brämer, W., Deiseroth, H.-J.:* Inorg. Chem. *17*, 875 (1978)
22. *Deiseroth, H.-J., Simon, A.:* unpublished

23. *Selte, K., Kjekshus, A.:* Acta Chem. Scand. *17*, 2560 (1963)
24. *Simon, A., Westerbeck, E.:* Z. anorg. allg. Chem. *428*, 187 (1977)
25. *Zickelbein, W.:* Thesis, Münster (1978)
26. *Simon, A.:* Z. anorg. allg. Chem. *422*, 208 (1976)
27. *Simon, A., Deiseroth, H.-J., Brämer, W.:* unpublished
28. *Simon, A., Deiseroth, H.-J., Westerbeck, E., Hillenkötter, B.:* Z. anorg. allg. Chem. *423*, 203 (1976)
29. *Tsai, K.-R., Harris, P.M., Lassettre, E.N.:* J. Phys. Chem. *60*, 338 (1956)
30. *Mack, K., Deiseroth, H.-J., Simon, A.:* unpublished
31. *Simon, A., Mack, C., Brämer, W.:* unpublished
32. Brämer, W.: Thesis, Münster (1974)
33. *Biltz, W.:* Raumchemie der festen Stoffe. Leipzig: Leopold Voss 1934
34. *Copley, J.R.D., Simon, A.:* unpublished.
35. *Richards, S., Kasper, J.S.:* Acta Cryst. *B25*, 237 (1969)
36. *Hoppe, R.:* Angew. Chem. *78*, 52 (1966) (Intern. Ed., *5*, 95)
37. *Hoppe, R.:* Angew. Chem. *82*, 7 (1970) (Intern. Ed., *9*, 25)
38. *Hoppe, R.:* Crystal structure and chemical bonding in inorganic chemistry. Rooymans, C.J.M., Rabenau, A. (eds.). North Holland Publ. Co. 1975, p. 127
39. *Hoppe, R., Meyer, G.:* private communication
40. *Ebbinghaus, G.:* Thesis, Stuttgart (1977)
41. *Stolz, H.J., Ebbinghaus, G.:* Verhandl. Dtsch. Phys. Ges. (VI) *12*, 309 (1977)
42. *Martin, T.P.:* J. Chem. Phys. *67*, 5207 (1977)
43. *Welch, D.O., Lazaretz, O.W., Dienes, G.J., Hatcher, R.D.:* Bull. Am. Phys. Soc. *22*, 394 (1977)
44. *Martin, T.P.:* J. Chem. Phys., in press
45. *Martin, T.P.:* Phys. Rev. *B15*, 4071 (1977)
46. *Martin, T.P., Stolz, H.J., Ebbinghaus, G., Simon, A.:* J. Chem. Phys., in press
47. *Burt, M.G., Heine, V.:* J. Phys. C: Solid State Phys. *11*, 961 (1978)
48. *Bauhofer, W.:* J. Phys. E: Scientif. Instr. *10*, 1212 (1977)
49. *Brauer, G.:* Z. anorg. Chem. *255*, 101 (1947)
50. *Dudgale, J.S., Phillips, D.:* Proc. R. Soc. *287A*, 381 (1965)
51. *Louie, S.G., Cohen, M.L.:* Phys. Rev. *B10*, 3237 (1974)
52. *Wittig, A.H.:* Phys. Rev. Lett. *24*, 812 (1970)
53. *Toulence, M.:* private communication
54. *Touzain, Ph.:* Bull. Soc. Chim. France *1973*, 4515
55. *Sinha, A.K.:* Phys. Rev. *B1*, 4541 (1970)
56. *Chen, H.S., Park, B.K.:* Acta Met. *21*, 395 (1973)
57. *Polk, D.E.:* Acta Met. *20*, 485 (1972)
58. *Nagel, S.R., Tauc, J.:* Phys. Rev. Lett. *35*, 380 (1975)
59. *Güntherodt, H.-J.:* Festkörperprobleme XVII, Adv. Solid State Physics. Treusch, J. (ed.). Braunschweig: Vieweg 1977, p. 25
60. *Hayes, T.M., Allen, J.W., Tauc, J., Giessen, B.C., Hanser, J.J.:* Phys. Rev. Lett. *40*, 1281 (1978)
61. *Nold, E., Steeb, S., Lamparter, P.:* Z. Naturforsch., in press
62. *De Boer, J.H., Broos, J., Emmens, H.:* Z. anorg. allg. Chem. *191*, 113 (1930)
63. *Mack, K., Simon, A.:* unpublished
64. *Momin, J., Boutry, G.-A.:* Phys. Rev. *B9*, 1309 (1974)
65. *Ebbinghaus, G., Braun, W., Simon, A.:* Z. Naturforsch. *31b*, 1219 (1976)
66. *Ebbinghaus, G., Simon, A.:* Submitted to Phys. Rev. *B*
67. *Davenport, J.W., cited by Helms, C.R., Spicer, W.E.:* Physics Lett. *57A*, 369 (1976)
68. *Coles, B.R., Caplin, D.A.:* The electronic structure of solids. London: Arnold 1976, pp. 44
69. *Smith, N.V., Spicer, W.E.:* Phys. Rev. *188*, 593 (1969)
70. *Barrie, A., Street, F.J.:* J. Electron Spectrosc. Relat. Phenom. *7*, 1 (1975)
71. *Eastman, D.E., Freeouf, J.L.:* Phys. Rev. Lett. *34*, 395 (1975)
72. *Kim, K.S.:* Phys. Rev. *B11*, 2177 (1975)

73. *Honig, J.M., Sinha, A.P.B., Wahnsiedler, W.E., Kuwamoto, H.:* Phys. Stat. Sol. (b) *73*, 651 (1976)
74. *Gregory, P.E., Chye, P., Sunami, H., Spicer, W.E.:* J. Appl. Phys. *46*, 3525 (1975)
75. *Peterssen, L. G., Karlsson, S. E.:* Proc. 5th Int. Conf. Vacuum Ultraviolet Radiation Physics, Vol. II (1977), p. 253
76. *Helms, C.R., Spicer, W.E.:* Phys. Rev. Lett. *28*, 565 (1972)
77. *Helms, C.R., Spicer, W.E.:* Phys. Rev. Lett. *32*, 228 (1974)
78. *Schön, G.:* Acta Chem. Scand. *27*, 2623 (1975)
79. *Kishi, K., Ikeda, S.:* J. Phys. Chem. *18*, 107 (1974)
80. *Johnson, O.:* Chemica Scripta *8*, 162 (1975)
81. *Schön, G.:* Surface Sci. *35*, 96 (1973)
82. *Oswald, R. G., Callcott, T.A.:* Phys. Rev. *B4*, 4122 (1971)
83. *Kullmann, H.-J.:* Staatsarbeit, Münster (1974)
84. *Simon, A., Brämer, W., Hillenkötter, B., Kullmann, H.J.:* Z. anorg. allg. Chem. *419*, 253 (1976)
85. *Helms, A., Klemm, W.:* Z. anorg. Chem. *242*, 34 (1939)
86. *Tsai, K.-R., Harris, P.M., Lassettre, E.N.:* J. Phys. Chem. *60*, 338 (1956)
87. *Kunz, C.:* Z. Physik *196*, 311 (1966)
88. *Ebbinghaus, G., Braun, W., Simon, A., Berresheim. K.:* Phys. Rev. Lett. *37*, 1770 (1976)
89. *Koller, L.R.:* J. Opt. Soc. Am. *19*, 135 (1929)

The Electronic Structure of Cobalt(II) Complexes with Schiff Bases and Related Ligands*)

Claude Daul, Carl Wilhelm Schläpfer and Alexander von Zelewsky

Institute of Inorganic and Analytical Chemistry,
University of Fribourg, CH-1700 Fribourg, Switzerland

Table of Contents

The electronic structure of low spin Co(II) complexes with *Schiff* base ligands is considered from an experimental and a theoretical point of view. EPR spectra can be explained in terms of a model which takes into account four electronically excited states, two doublets and two quartets, respectively. These excited states lie all very close to the ground state. Other experimental results are more difficult to interpret than the EPR spectra, but are essentially in agreement with the scheme proposed for the state energies. Several theoretical methods to calculate the state energies are compared.

*) This study was supported by the Swiss National Science Foundation

I. Introduction

Schiff bases derived from aldehydes or ketones form complexes with most of the transition metals (*41*). The doubly bonded nitrogen atom causes rather strong ligand fields, i.e. a large splitting of the d-orbital energies and, consequently, a preferential occurrence of low-spin configurations is observed in such complexes (*42*).

Due to the Jahn-Teller effect, cobalt(II) as a d^7 ion is expected to be unstable in an octahedral environment if the ligand field is strong enough to cause spin pairing. Consequently, six-fold coordination is always accompanied by a strong deviation from octahedral symmetry as demonstrated by $Co(CNC_6H_5)_6^{2+}$, the only low spin Co(II) complex with six identical ligands (*45*). Low spin Co(II) complexes often have a five- or four-fold coordinated metal, the latter in a more or less planar arrangement of the donor atoms. Despite the fact that such complexes have been known for a long time (*62*) their detailed electronic structure has been investigated intensively only during the last few years (*20*). The oxygen-binding properties of some of these compounds were a particular stimulus to this kind of work (*8*).

As with many planar complexes of other metals, there has been some dispute about the exact ordering of energy levels. One of the main subjects of this article is to try to clarify these questions for Co(II) complexes with *Schiff* bases and related compounds as ligands. We shall exclude from our discussion compounds of high symmetry, e.g. porphyrins, on which review articles have been written (*52*) and also complexes with "tripod" ligands (*17*). In general we shall restrict our discussion to complexes with a distinct deviation from axial symmetry as is the case for complexes formed by coordination of *Schiff* bases.

II. Structure

The complexes considered here can be represented by the general formula:

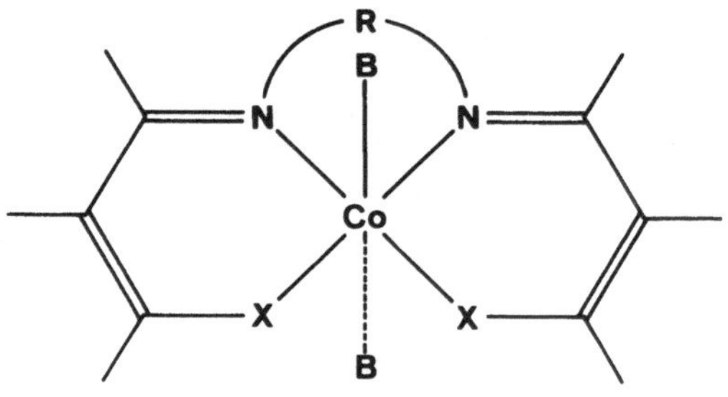

The in-plane ligand can be characterized by two non-saturated six-membered chelate rings connected by the bridge R. The resulting cis-$N_2X_2Co(II)$ complex generally carries no net charge. Spectroscopic and magnetic properties of a number of complexes considered in more detail are given in Appendix 1 together with the abbreviations used in this article.

The *Schiff* base complexes of M^{2+} show a great variability in their structures. This is demonstrated by Cu(acacen), which crystallizes in three different forms, as the anhydride (*33*), as the hemihydrate (*18*) and the monohydrate (*34*). A similar behaviour is shown by Ni(acacen), which was used as matrix for Co(acacen) in single crystal EPR work. Mixed crystals grown from the vapour phase are orthorhombic, space group $Pna2_1$ (*12*). Crystals grown from a acetone solution are monoclinic (*5*) and contain 1/2 H_2O.

Co(salen) has at least two different types of crystal structure, depending on the solvent of crystallization. A monomeric structure containing one solvent molecule per complex unit is obtained from chloroform (*65*). The cobalt ion is surrounded by the 2N + 2O donor atoms in an essentially planar arrangement. Crystallization from other solvents, e.g. acetone, yield a dimeric structure with oxygen bridges (*24*). Furthermore, most of the complexes easily add a fifth or a sixth ligand thereby forming a penta- or hexacoordinated species, e.g. Co(salen)py which can be crystallized from pyridine. In Co(salen)py the axial Co–N distance is 2.10 Å and the in-plane Co–N distance is 1.90 Å (*13*). Recently the formation of numerous adducts of Co(saphen) with O, S, Se, Te, N and P donors has been investigated (*32*).

The behaviour of the Ni complexes is much simpler than that of the corresponding Co compounds. Generally there appears to be little or no axial coordination. The tendency to crystallize with solvent molecules is however observed in some cases. Ni(salen) crystallizes (e.g. from $CHCl_3$) with one solvent molecule per unit cell, as does Co(salen) (*51*).

As will be seen later on, the electronic structure of the cobalt complexes is extremely sensitive to axial perturbation. It is therefore imperative to pay much atten-

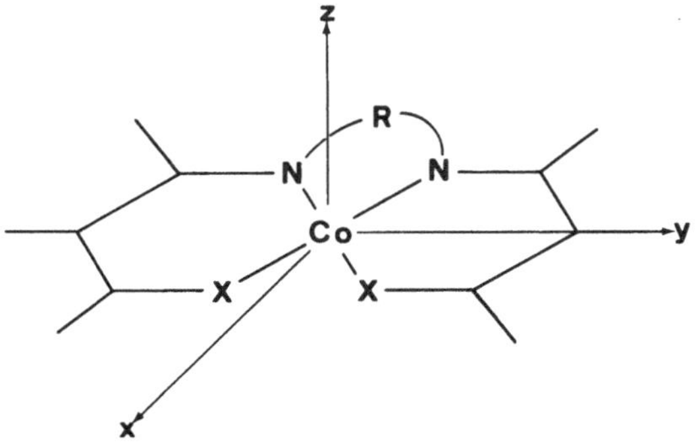

Fig. 1. Cartesian coordinate system used for Co(II) complexes

tion to the environment of a complex, whether it is studied in solution, in a neat crystal, or in magnetically diluted crystals. The influence of axial coordination will be discussed in detail in Chapter III. 2. of this article.

Another important point to which not always sufficient consideration has been given, despite its trivial nature, is the choice of the axis system. Most planar complexes (without axial ligands) have exact or approximate C_{2v} symmetry. The five-fold coordinated complexes can be treated in the same point group as the planar species, if the concept of holoedrized symmetry is applied (*43*). The coordinate system most often used to discuss octahedral complexes, in which the ligand atoms lie on the three axes, is therefore not practical. Instead of this, the *in-plane* Cartesian axes should be chosen as shown in Figure 1.

Almost all character tables are constructed by convention in such a way that the z axis coincides with the axis of highest symmetry. In C_{2v}, therefore the z axis should be chosen *in*-plane, which is contrary to custom in coordination chemistry. We have retained the z direction perpendicular to the molecular plane (*75*). The character table has to be relabelled, as shown in Table 1. *Hitchman* (*37*) has used identical designations.

Table 1. Character table for the symmetry group $C_{2v}(x)$

	E	$C_2(x)$	$\sigma(xz)$	$\sigma(xy)$		
A_1	1	1	1	1	x	d_{z^2}, $d_{x^2-y^2}$
A_2	1	1	−1	−1	R_x	d_{yz}
B_1	1	−1	1	−1	z, R_y	d_{xz}
B_2	1	−1	−1	1	y, R_z	d_{xy}

III. EPR Spectroscopic Investigations

1. Planar Complexes, Coordination Number Four

Figure 2 shows the Q band (35 GHz) powder spectrum of Co(acacen) doped into Ni(acacen) as a typical example of the EPR spectra of planar, four-fold coordinated, low-spin Co(II) *Schiff* base complexes. The structure, due to the three principal values of the g and A tensors is nicely resolved. Both tensors are clearly orthorhombic, with one large and two similar, smaller principal values. Powder spectra yield however no information on the orientation of the tensors with respect to the molecular frame. This information is extremely important in complexes having low molecu-

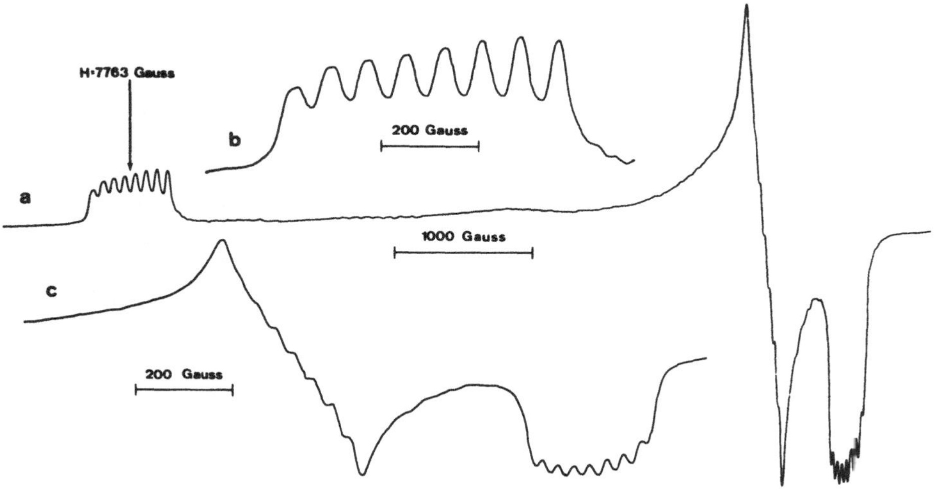

Fig. 2. EPR spectrum of Co(acacen) doped into Ni(acacen), polycrist. sample, T = 120 K, ν = 35.1 GHz.
a) complete spectrum; b) expanded low field part; c) expanded high field part

lar symmetry (no C_n axis with $n \geqslant 3$), because the tensors characterizing the Spin-Hamiltonian will show in general no special symmetry properties. This means that each of these tensors has to be described not only by three principal values but also by three directional parameters, which are determined by the ground state wave function. It is therefore possible that complexes, which are structurally similar, show very different anisotropies of the tensors. One of the most striking examples are the complexes Cu(salen) and Co(salen). Both complexes have doublet ground states, are closely related structurally and can be doped into Ni(salen), which is diamagnetic. In the coordinate system defined in Chapter II the principal values of Cu(salen) are (*67*) $g_x = 2.049$; $g_y = 2.046$ and $g_z = 2.192$. The g tensor accordingly shows nearly axial symmetry about the z axis which is *perpendicular* to the molecular plane.

The investigation of a single crystal (*29, 75*) of Ni(salen) grown from acetone and doped with Co(salen) *and* Cu(salen) showed unambiguously that the large principal component of the g tensor is g_x. The principal values (*75*) $g_x = 3.805$, $g_y = 1.65$ and $g_z = 1.74$ show nearly axial symmetry but with the "symmetry" axis lying *in* the molecular plane. A similar observation was reported earlier for complexes with orthorhombic ligand fields containing sulfur ligands (*30, 46*). Planar Co(II) complexes with higher symmetry, e.g. the phthalocyanines (*2*) behave very differently; they have z as symmetry axis of the g tensor, reflecting the D_{4h} symmetry of the ligand field.

This large in-plane anisotropy which was first observed for Co(salen) (*75*) was later confirmed by detailed single crystal measurements (*47, 76, 44, 12*) on related compounds. Another, less rigorous method of obtaining directional information is the measurement of EPR spectra in frozen nematic phases. It confirms the results of the single crystal measurements (*40, 70, 76*). Based on this ample evidence, it is reasonable to assume that the orientations of the g and A tensors in all related cases is the same, even if only powder spectra are known.

In this context the results of one of the single crystal measurements deserve special attention. It has been claimed that in Co(acacen) diluted into Ni(acacen), the principal axes of the g and A tensors are oriented in a fashion which is not parallel to the axis system of Figure 1, but rather rotated approximately 45° about the z axis lying almost parallel to the Co ligand bond (*12*). The other special features of EPR spectra of planar complexes of this type, particularly the in-plane situation of the largest g axis (g_1 = 3.26) and the approximative axial symmetry (g_2 = 2.00; g_3 = 1.88) are preserved. A detailed reconsideration of the data makes a different assignment with g_1 lying parallel to the approx. two-fold axis x more plausible. This result is obtained if the g values of the spectra of the two magnetically non-equivalent sites are interchanged.

The large deviations of g_x from the g value of the free electron indicates a strong mixing of the ground state with low lying excited states by spin-orbit coupling. The energy separation of these states is of the order of the spin-orbit coupling constant $\lambda_{Co} \approx -400\ cm^{-1}$. This mixing is much larger than in analogous Cu(II) complexes. This fact is not surprising if one considers the splitting of the five d orbitals in a planar complex shown in Fig. 3.

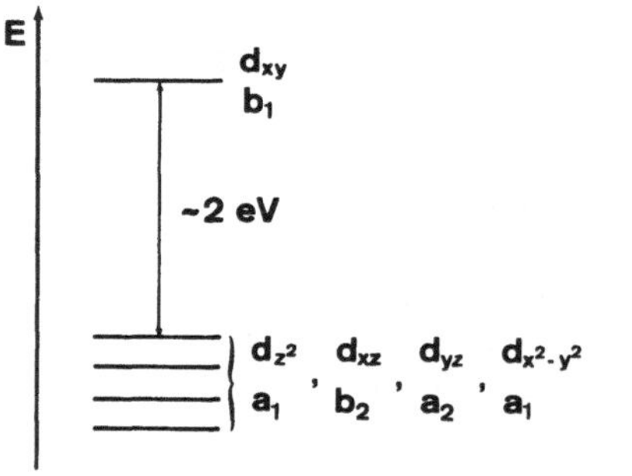

Fig. 3. d orbital energy scheme for planar complexes

One of the d orbitals, d_{xy} in the coordinate system shown in Figure 1, is strongly destabilized by σ interaction with the ligand. The energy difference between this orbital and the other d orbitals, which are close together in energy, is of the order of 2 eV. In the Cu(II) complexes (d^9), the single hole is in the d_{xy} orbital. In all excited states an electron is promoted from one of the four other orbitals to the d_{xy} orbital. In the d^9 system where no electron interaction has to be considered, the energies of the excited states are given by the orbital energies. Consequently, they are in the order of 2 eV, much larger than the spin-orbit coupling constant $\lambda_{Cu^{2+}} = -820\ cm^{-1}$. Hence the mixing of these states to the ground state by LS coupling is small, and the EPR spectra can be interpreted by perturbation theory in terms of orbital energies (*49*). There is general agreement that this orbital is not occupied in Co(II) complexes, which is also confirmed by the fact that d_{xy} is the empty orbital in diamagnetic Ni(II) complexes. The unpaired electron resides in the Co(II) complex, therefore in one of the four low-lying d orbitals. Obviously there will be low lying excited states, produced by the excitation of electrons within this set, which can be mixed into the ground-state by LS coupling. This has to be kept in mind for the interpretation of the EPR parameters, and the perturbation theory has to be applied with the necessary precautions. Even though it is tempting to treat the system under these conditions as a one-hole system with a reduced set of four 3d orbitals, one has to stress that the energies obtained are always state energies and not d-orbital energies. The latter are obtained only after allowing for interelectronic interactions. Despite this, in the following we will label the states to indicate only the orbital of the one hole, within the reduced set, assuming d_{xy} to be empty.

A detailed study of the different interactions of the four possible states by LS coupling clearly indicates that only $|yz, {}^2A_2\rangle$ or $|z^2, {}^2A_1\rangle$ as ground states, can explain the large value of g_x. The large anistropy and the orientation of the principal axes can, indeed, be understood in a first approximation in the very simple $|yz, {}^2A_2\rangle / |z^2, {}^2A_1\rangle$ *crossover* model. In this model the two states derived from $|yz, {}^2A_2\rangle$ and $|z^2, {}^2A_1\rangle$ are assumed to be the lowest doublets, changing their relative energies. As we shall see later, the state energy is very sensitive to any change in σ interaction, especially to axial perturbations. We therefore chose a diagram in which the energy of $|yz, {}^2A_2\rangle$ is constant and that of $|z^2, {}^2A_1\rangle$ varies linearly (Fig. 4). The lower of the two states will be the ground state.

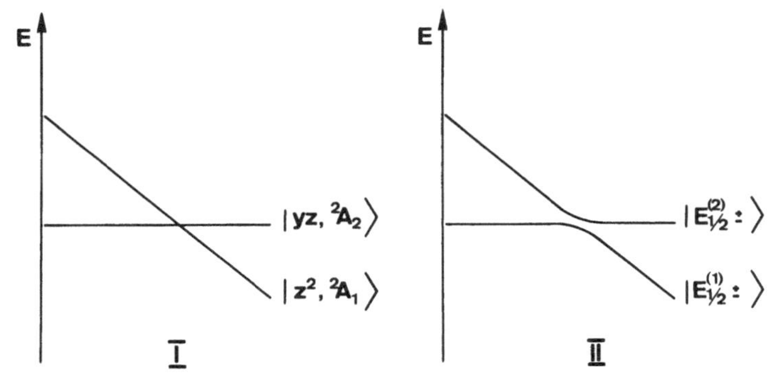

Fig. 4. Two state *crossover* model; (I) no spin-orbit coupling, symmetry labels according to $C_{2v}(x)$; (II) spin orbit coupling included, symmetry labels according to $C^*_{2v}(x)$

The two states are the only ones to be considered in this approximation. In the $C_{2v}(x)$ symmetry, $|yz\rangle$ transforms as a_2, and $|z^2\rangle$ according to a_1 (also $d_{x^2-y^2}$ transforms according to a_1, and configuration interaction of the corresponding states is therefore possible). $|yz,{}^2A_2\rangle$ and $|z^2,{}^2A_1\rangle$ transform according to different representations and a crossing of the two states is therefore possible if ligand field parameters change (Fig. 4, I). If spin-orbit coupling is included in the Hamiltonian, crossing of the two states is no longer possible because they then transform according to the same irreducible representation of the *double group* C^*_{2v} (x). We therefore introduce new functions:

$$
\begin{aligned}
|E^{(1)}_{1/2}\ \pm\rangle &= a\,|yz,{}^2A_2\ \pm\rangle + ib\,|z^2,{}^2A_1\ \mp\rangle \qquad \text{ground state}\\
|E^{(2)}_{1/2}\ \pm\rangle &= b\,|yz,{}^2A_2\ \pm\rangle - ia\,|z^2,{}^2A_1\ \mp\rangle \qquad \text{excited state}
\end{aligned}
\tag{1}
$$

a and b are the mixing coefficients due to spin-orbit coupling. The exact energies and eigenfunctions within this model can be given in terms of en energy parameter Δ (energy difference between $|z^2,{}^2A_1\rangle$ and $|yz,{}^2A_2\rangle$) and the spin-orbit coupling constant λ, (eq. 2).

$$
\begin{aligned}
a &= \cos\theta \qquad \text{where } \theta = -1/2\tan^{-1}\frac{\lambda\sqrt{3}}{\Delta}\\
b &= \sin\theta \qquad 0 \leqslant \theta \leqslant \pi/2
\end{aligned}
\tag{2}
$$

The two levels can approach each other to $\lambda\sqrt{3}/2$ in which case $a = b = \sqrt{2}/2$. We shall speak of a $|yz,{}^2A_2\rangle$ or $|z^2,{}^2A_1\rangle$ ground state if $b > a$ or $a > b$, respectively. The equations for the g and A values in this scheme are:

$$
\begin{aligned}
g_x &= 2(1 + 2\sqrt{3}\,ab) & A_x &= P\{a^2(-\kappa - \tfrac{4}{7}) + b^2(-\kappa - \tfrac{2}{7}) + 4\,ab\sqrt{3}\}\\
g_y &= 2(a^2 - b^2) & A_y &= P\{a^2(-\kappa + \tfrac{2}{7}) - b^2(-\kappa - \tfrac{2}{7}) - \tfrac{2}{7}\,ab\sqrt{3}\}\\
g_z &= 2(a^2 - b^2) & A_z &= P\{a^2(-\kappa + \tfrac{2}{7}) - b^2(-\kappa + \tfrac{2}{7}) + \tfrac{2}{7}\,ab\sqrt{3}\}
\end{aligned}
\tag{3}
$$

The symbols have their usual significance (*31*). The g values are given as functions of the mixing parameter a^2 in Figure 5, I and as functions of Δ/λ in Figure 5, II.

The features of this model are:

a) g_x is the largest g component. It can assume very large values ($g_x^{max} = 5.46$).
b) g_y and g_z are both $\leqslant 2$.
c) All g values are "symmetric" in Figure 5 with respect to $a^2 = b^2 = 0.5$ ($\Delta/\lambda = 0$).
d) $g_y = g_z$, which means there is axial symmetry about the x axis.

The first two points are in agreement with the experimental facts. c) shows that this model can approximately give the distance from the "crossover" point, but it does not allow us to distinguish whether a particular complex has a $|z^2,{}^2A_1\rangle$ or a $|yz,{}^2A_2\rangle$ ground state. Eq. 3 shows that the A values are not symmetric with respect

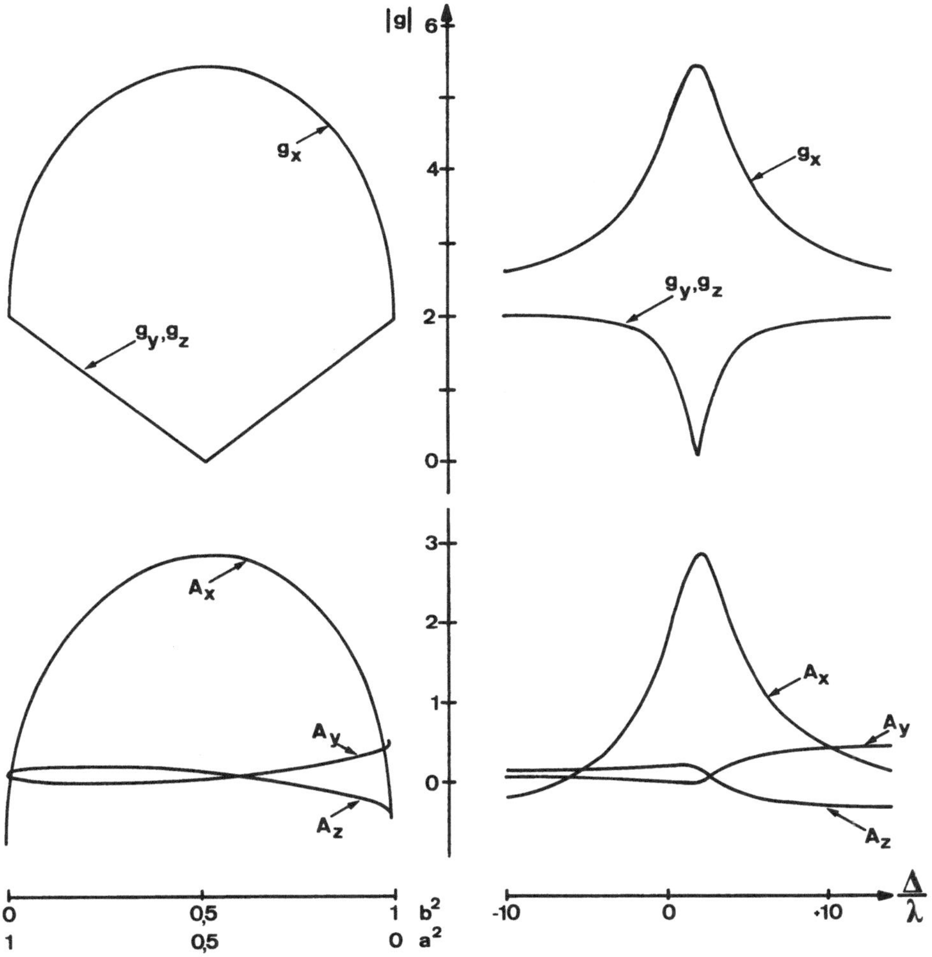

Fig. 5. EPR parameters in the two state *crossover* model; (I) as functions of a^2; (II) as functions of Δ/λ

to $a^2 = b^2 = 1/2$. Therefore, they can be used to distinguish between the two ground states (*29*). Accurate values are available for all g values and A_x. A_y, and A_z splittings are difficult to measure because they are rather small and often not well resolved. It is therefore most reasonable to correlate g_x and A_x. A representation of this correlation is given in Figure 6.

All values obtained from complexes in Ni matrices and nematic phases lie on the branch of the curve which belongs to $a^2 > 1/2$, whereas the base adducts discussed in

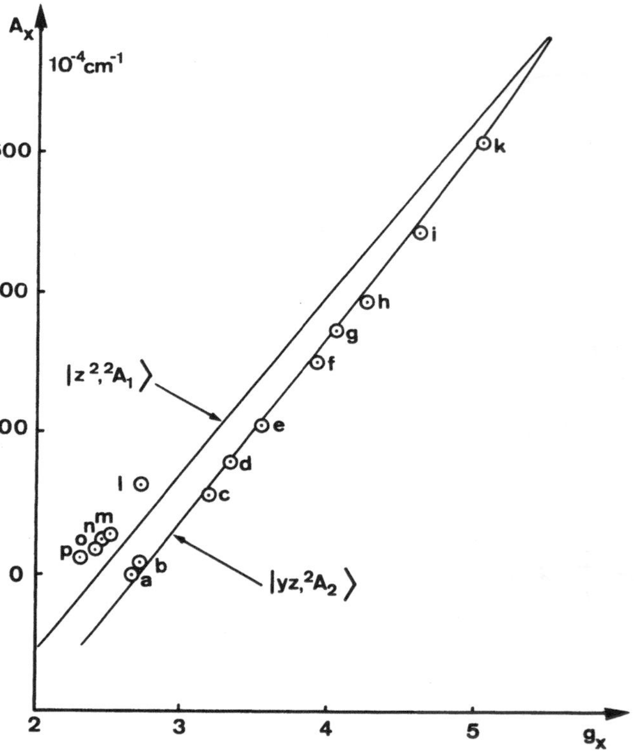

Fig. 6. Correlation diagram of A_x vs. g_x. The solid line represents the calculated function in the two state *crossover* model ($P = 230 \cdot 10^{-4}$ cm^{-1}, $\kappa = 0.15$).
Experimental values are given for: a) Co(amben); b) Co(amben) n.p.; c) Co(acacen); d) Co(bzacacen); e) Co(CF_3acacen); f) Co(salen); g) Co(salen) in Pd(salen); h) Co($a_2$en) in Pd($a_2$en); i) Co($a_2$phen) in Pd($a_2$phen); k) Co($a_2$phen); l) [Co(salen)]$_2$; m) Co(saphen)thf; n) Co(saphen)N(C_2H_5)$_3$; o) Co(saphen)py; p) Co(saphen)py$_2$

III. 2 lie close to the other branch. The deviation is due to low-lying quartet states which influence the A_x values strongly (vide infra).

The experimental deviation of the g tensor from axial symmetry about the x axis cannot be explained in this model, and a more refined model with enlarged basis set has to be used.

It is sufficient to carry out calculations in the same approximation as before, including a third state, namely $|xz, {}^2B_1\rangle$. If the model is extended in this way, assuming a constant energy separation of $|yz, {}^2A_2\rangle$ and $|xz, {}^2B_1\rangle$, a result indicated in Figure 7 is obtained.

Independently of the assumption of the magnitude of this separation, the qualitative result is $g_z > g_y$ for a $|yz, {}^2A_2\rangle$ ground state and $g_y > g_z$ for the $|z^2, {}^2A_1\rangle$

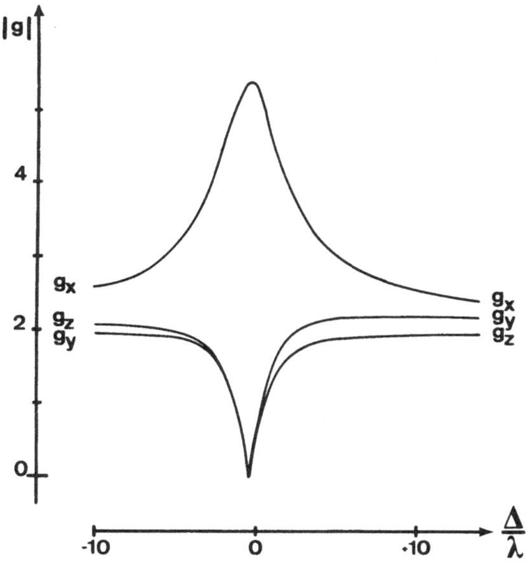

Fig. 7. g_x, g_y and g_z in the extended *crossover* model including the orthorhombic splitting as a function of Δ/λ (Δ orthorhombic = 12 λ)

ground state. On the basis of this simple model, it is therefore possible to deduce the ground state electronic configuration merely from the sequence of the three g values (*50*). The orthorhombic splitting is due to a difference in interaction of d_{xz} and d_{yz} with ligand π orbitals (cf. 7.1). This simple criterion for the assignment of the ground state is in agreement with the independent conclusions drawn from Figure 6 and more rigorous models (cf. VII. 3). Using this energy scheme the g values of the planar complexes can be explained without exception.

If the interpretation of the g tensor in planar complexes is quite straightforward once the correct energy diagram has been chosen, the same is not the case for the hyperfine tensor, with the exception of A_x. As several authors have pointed out, a correct interpretation of the A tensor is not possible without considering explicitly low-lying quartet states of the complexes (*50, 37*).

2. Complexes with Axially Coordinated Bases

The Q-band EPR spectrum of Co(saphen)$P(OC_2H_5)_3$ is shown in Figure 8 as a typical example for this class of complexes (cf. Appendix).

Obviously axial coordination has a profound influence on the EPR spectra of the cobalt complexes. The spectra show a smaller anisotropy which points to a higher en-

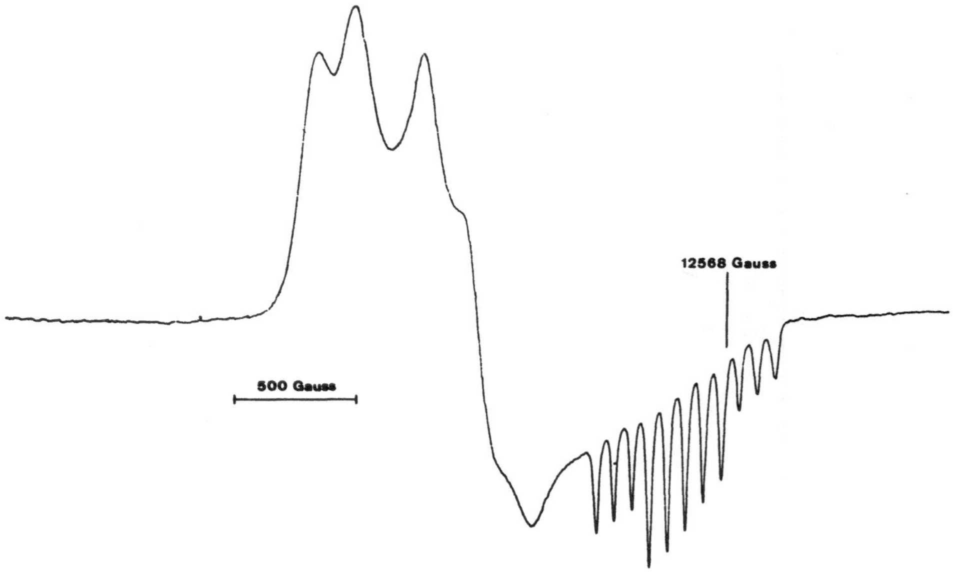

Fig. 8. EPR spectrum of $Co(saphen)P(OC_2H_5)_3$ in glassy solution, $T = 120\,K$, $\nu = 35.1\,GHz$

ergy of the first excited doublet, compared with the planar complexes. Only the signal due to the lowest g component shows cobalt-hyperfine-splitting, the other components of the hyperfine tensor can in general be determined by curve fitting. All of these measurements have been carried out in frozen solutions or nematic phases (*40*), since single crystal studies are not feasible because suitable host lattices are lacking. In some cases, it can be decided unambiguously whether one or two bases are coordinated from the hyperfine structure of the axial ligands.

The magnitude of this superhyperfine structure ($A^N \approx 15 \cdot 10^{-4}$ cm, $A^P \approx 70 \cdot 10^{-4}\,cm^{-1}$, $A^{Te} \approx 187 \cdot 10^{-4}\,cm^{-1}$ (*73, 32*)) is characteristic for a σ interaction of the unpaired electron with the axial ligands. Hence, the assignment of the ground state as a $|z^2, {}^2A_1\rangle$ state is certain, despite the fact that in frozen solutions the direction of the tensor axes cannot be determined unambiguously. There is general agreement that the $|z^2, {}^2A_1\rangle$ ground state prevails in all adducts, and it is indeed also found in a Co(II) complex with a corrin ligand with axial coordination (*74*), and in similar complexes with D_{4h} symmetry (*72*). From a simple MO consideration, this ground state is very plausible. Axial coordination *destabilizes* $|z^2, {}^2A_1\rangle$, which becomes more σ antibonding than in the planar complexes, causing a change in the ground state. In either of the two models (Figs. 5 and 7), the base adducts are situated to the right, having a large and positive value for $\Delta = E\,|z^2, {}^2A_1\rangle - E\,|yz, {}^2A_2\rangle$. As expected Δ is larger in the diadducts than in the monoadducts. The g values can be interpreted in a good approximation in the same simple model as described in Chapter III. 1. (*32*).

There are indications that planar Co(II) complexes in coordinating solvents show high-spin/low-spin equilibria (*16, 48*). The high-spin states are stabilized by strong axial interaction (*38*). This is in agreement with the results of semi-empirical calculations showing that, in the base adducts, some quartet levels are even closer to the ground state than in the four-coordinated planar complexes. Their influence in EPR spectroscopy is mainly on the hyperfine parameters. This will be discussed in Chapter VII.

It should be noted that the inactive form of Co(salen) fits very well into the series of axially coordinated complexes (*37*). This result has been observed, however, from powder measurements, and a definite assignment of the axes is therefore not possible.

Some of the complexes have also been measured in frozen nematic phases (cf. Appendix). The different absolute values of the largest g components suggest an axial perturbation by matrix molecules. Whether in the nematic phases, five- or four-fold coordinated complexes are present is not completely certain. All assignments (*40, 70, 76*) lead to a maximum g component *in* the molecular plane. In those cases in which a definite assignment of the orientation of the molecule can be made, the large g component turned out to be g_x. An assignment is only possible if the complex has a distinct rectangular shape, and it is much less certain in Co(acacen), which has an approximately circular molecular shape with no preferential orientation with respect to the x or y axes. No additional information is available from the analogous copper compounds, since those complexes are essentially isotropic in the molecular plane. Nematic phase measurements can contribute to the elucidation of EPR parameters in planar paramagnetic complexes, but the information is in some cases limited to the assignment of one axis only.

IV. Magnetic Susceptibilities

Bulk magnetic susceptibilities of Co(salen) and similar compounds have been measured as a function of temperature in the range 100 K to 400 K by *Calvin* et al. (*8*). In several cases, room-temperature susceptibility measurements have confirmed the low-spin ground state of these complexes.

More recently, low-temperature and single-crystal measurements have been carried out (*55, 56, 11*). Most measurements have been made with neat crystals where the coordination number is five, e.g. Co(salen)py, $[Co(salen)]_2$ and $[Co(tsalen)]_2$. According to the EPR results, this would most probably mean a $|z^2, {}^2A_1\rangle$ ground state, which is also deduced from the susceptibility measurements.

Interpretation of measurements on presumably monomeric complexes are not in agreement with EPR results. As the magnetic properties of these complexes are strongly dependent upon the surroundings (cf. Chapter III), comparisons of the susceptibility measurements with spectroscopic data obtained from magnetically diluted samples are difficult. The theoretical treatment given in (*56*) is unfortunately not directly comparable with the interpretation of the EPR results because the state functions used were derived directly from octahedral strong field configurations. This is not adequate for planar complexes, in which deviation from octahedral symmetry gives extremely important effects on the magnetic properties.

The most important result from the magnetic measurements is the susceptibility maximum in $[Co(salen)]_2$ at low temperatures, indicating a coupling between the two unpaired electrons in the dimer. The low temperature susceptibility data reported by the two groups are not in complete agreement, yielding $J = -20.0\,cm^{-1}$ (*11*) and $J = -40\,cm^{-1}$ (*56*), respectively. There is, however, a qualitative agreement about the sign of the exchange interaction, and the interpretation of the data in terms of a superexchange mechanism is probably correct.

The deviations of the susceptibilities of $[Co(salen)]_2$ at higher temperatures, from models using only doublet state contributions, might be interpreted as being due to thermal interaction with low-lying quartet states. These deviations are even stronger in Co(salen)py where quartet states are probably even closer to the ground state (*56*).

V. NMR Spectra

The NMR spectra of planar Co(II) complexes have been measured by two groups (*53, 54, 68, 69*) in order to obtain direct experimental information on the metal–ligand interactions from the isotropic chemical shifts. The spectra in noncoordinating solvents ($CDCl_3$) clearly indicate a transfer of α spin density into the HOMO of the ligand π system. This is interpreted in different ways. *Srivanavit* and *Brown* (*68, 69*) assume a $|z^2, {}^2A_1\rangle$ ground state. It is then difficult to explain the considerable transfer of spin density from the d_{z^2} orbital into the ligand π orbital. On the other hand, *Migita* et al. (*53*) assume a $|yz, {}^2A_2\rangle$ ground state, which explains quite naturally the transfer of spin-density and is in agreement with the EPR results.

The NMR results from measurements in coordinating solvents like pyridine (*54*) or dmso (*69*) can be interpreted as being due to a $|z^2, {}^2A_1\rangle$ ground state, which is again in agreement with the conclusions of the other chapters. Spin-density is transferred through σ and π interactions from the metal ion onto the ligand.

The temperature dependence of the isotropic shifts deviates from the Curie-Weiss law. Both groups attribute this observation to the influence of low-lying quartet states, which couple to the ground state via spin-orbit interaction.

VI. Electronic Spectra

The EPR parameters and the state energies, estimated by any theoretical method (Chapter VII), indicate that low spin Co(II) complexes have a number of excited doublet and quartet states with energies less than 10 kK above the ground state. The transitions into these states have mainly d-d character, and give rise to weak absorption bands. The measurement of the d-d spectra of these complexes, which are generally not very soluble in inert solvents, is therefore experimentally difficult and the interpretation of the spectra poses the following problem. The energy of the d-d transitions is of the order of the vibrational energies. As a consequence, the B.O. approximation might break down, and vibronic mixing has to be considered.

Figure 9 shows the absorption spectrum of Co(acacen), dissolved in $CHCl_3$, from 3.5 to 16 kK. At energies higher than 16 kK strong absorptions due to interligand or CT transitions obscure any other d-d transition. To obtain information on d-d transitions in this region, complexes of optically active ligands have been prepared (*36, 7, 70*), and their CD spectra measured. However, these are complicated and their interpretation highly speculative. The absorption bands due to transitions into states with energies lower than 3.5 kK are obscured by strong vibrational absorptions and, so far, have never been observed.

The spectrum in Figure 9 is characteristic for four-fold coordinated Co(II) complexes, with a planar N_2O_2 coordination sphere independent of the detailed structure of the ligand (*57, 70, 37*). This shows, in contrast to the EPR spectra, that the ab-

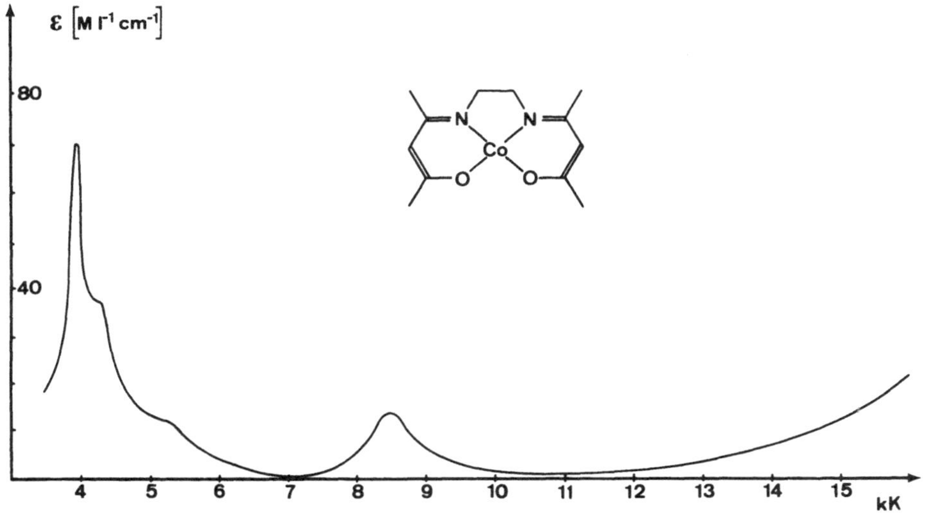

Fig. 9. Absorption spectrum of Co(acacen) in $CHCl_3$, T = 300 K

sorption spectra are insensitive to small changes in metal-ligand π interaction. If, however, the two O ligand atoms in the coordination sphere are replaced by NH (*70*) or S (*66*), the spectrum changes, reflecting a change in the relative d-orbital energies, which is not surprising because in this case it is the dominant σ interaction which is affected. In the near infrared no solution spectra of five-fold coordinated complexes have been measured. Reflectance spectra of [Co(salen)]$_2$ and Co(salen)py, however, have been reported (*7, 37*). Their appearance is different compared to the spectrum in Figure 9. Two peaks of similar intensity are observed at 6.0 and 11.5 kK in [Co(salen)]$_2$, and three peaks in Co(salen)py at 6.0, 9.0 and 11.5 kK. Other very weak bands are tentatively assigned to vibrational overtones.

The interpretation of the d-d absorptions is not completely clear. Several different assignments have been given (*7, 70, 14, 37*). It is difficult to directly compare the results from electronic absorption spectra with the conclusions drawn from EPR spectroscopy, because the former involve states with energies above 4 kK, much larger than $\lambda_{Co} \approx -400\,cm^{-1}$. Hence, these states only weakly influence the EPR parameters. The energies of the states which dominate the EPR parameters are below 4 kK, the low-energy limit of the observed electronic transitions. Reviewing the assignments critically, the most recent, put forward by *Hitchman* (*37, 38*), can be considered as the most probable. This assignment (Table 2) has been obtained from an estimate of state energies, using independently obtained semi-empirical parameters (cf. Chapter VII) which allow a unified interpretation of EPR and absorption spectra of four- and five-fold coordinated complexes (cf. Chapter VII).

Table 2. Assignment of the d-d transitions according to (*37*). kK(ϵ)

	Increasing axial perturbation Co(salen) ($CHCl_3$)	 [Co(salen)]$_2$ refl. (77 K)	 Co(salen)py refl. (77 K)
$\lvert yz, {}^2A_2\rangle \rightarrow \lvert xz, {}^2B_1\rangle$	3.90 (65)	–	–
$\rightarrow \lvert x^2-y^2, {}^2A_1\rangle$	8.30 (16)	–	–
$\lvert z^2, {}^2A_1\rangle \rightarrow \lvert yz, {}^2A_2\rangle$	–	–	6.00
$\rightarrow \lvert xz, {}^2B_1\rangle$	–	6.10	9.00
$\rightarrow \lvert x^2-y^2, {}^2A_1\rangle$	–	11.50	~ 16.00

VII. Theoretical Investigations

1. MO Descriptions

MO calculations at different levels of sophistication have been carried out (*21, 28, 23, 37*). The results of these calculations are essentially consistent with the conclusions drawn from EPR measurements (Chapter III). There is complete agreement about the ground state of five-fold coordinated complexes, which is invariably $|z^2, {}^2A_1\rangle$. For the four-fold coordinated complexes, all semi-empirical calculations predict $|yz, {}^2A_2\rangle$ ground state, whereas an *ab initio* minimal basis set calculation (*23*) favours the $|z^2, {}^2A_1\rangle$ ground state in the case of Co(acacen).

We shall discuss these calculations starting with the most simple *Angular Overlap Model* (AOM).

a) Angular Overlap Model

Because reliable data from electronic spectra of the Co(II) complexes are lacking, bonding parameters have to be taken from other sources. As pointed out by *Hitchman* (*37*), it is possible to determine the parameters of the ligand field of *Schiff* bases, using d orbital energies derived from the analogous Cu(II) complexes. Since the EPR spectra of Cu(II) *Schiff* base complexes indicate that the departure of the ligand field from axial symmetry is relatively small, the bonding parameters have been derived assuming axial symmetry, with the orthorhombic component being added subsequently as a perturbation. The AOM yields the following expressions for the ligand field splitting of the d orbitals and the crystal field parameters (*37*).

$$
\begin{aligned}
E(d_{xy}) &= 3\,e_\sigma(xy)\\
E(d_{z^2}) &= e_\sigma(xy) + e_\sigma(z)\\
E(d_{x^2-y^2}) &= 4\,e_{\pi\perp}(xy)\\
E(d_{xz}, d_{yz}) &= 2\,e_{\pi\parallel}(xy) + e_\pi(z) \qquad (4)\\
D_q &= \{3\,e_\sigma(xy) - 4\,e_{\pi\perp}(xy)\}/10\\
D_s &= \{e_\sigma(z) - 2\,e_\sigma(xy) + e_\pi(z) + 2\,e_{\pi\parallel}(xy) - 4\,e_{\pi\perp}(xy)\}/7\\
D_t &= \{3\,(e_\sigma(z) - 2\,e_\sigma(xy)) - 4\,(e_\pi(z) + 2\,e_{\pi\parallel}(xy) - 4\,e_{\pi\perp}(xy))\}/35
\end{aligned}
$$

Here, $e_\sigma(xy)$ represents the energy by which the d orbitals are raised upon interaction with one ligand donor atom in the xy plane, while $e_\sigma(z)$ represents the σ interaction of the axial ligand (this being zero when a four-coordinated complex is considered). The π interaction with the *Schiff* base is separated into two components, $e_{\pi\perp}(xy)$ and $e_{\pi\|}(xy)$, representing the interaction in- and out-of-plane, respectively; $e_\pi(z)$ denotes the π interaction of an axial ligand symmetrical about the bond axis. It is to be noted that, in these expressions, the parameters represent the arithmetic mean of the perturbations due to the oxygen and nitrogen atoms of the *Schiff* base, following the concept of "holohedrized symmetry" (*43*).

The variation in metal interaction on going from a Cu(II) to a Co(II) complex with an identical structure will be affected by an increase in the overlap integral, and by a greater separation between the coulomb integrals of the ligand and metal. These two effects should approximately balance. Using this procedure, *Hitchman* (*37*) obtained the ligand field bonding parameters of various Co(II) *Schiff* base complexes, as given in Table 3. Substitution of these paramaters into Eq. 4 yields the d orbital energies. The effect of the rhombic component of the ligand field was included by adding the terms:

$$\langle d_{yz} | V_{\text{ligand field}} | d_{yz} \rangle = 2\,500\ \text{cm}^{-1}$$

$$\langle d_{xz} | V_{\text{ligand field}} | d_{xz} \rangle = -\,2\,500\ \text{cm}^{-1}.$$

Table 3. Ligand bonding parameters of various Schiff base complexes in kK

Complex	$e_\sigma(xy)$	$e_{\pi\perp}(xy)$	$e_{\pi\|}(xy)$	$e_\sigma(z)$	$e_\pi(z)$
Co(salen)	11.666	4.036	6.631	–	–
$[Co(salen)]_2$	10.431	3.293	5.410	3.008	0.437
Co(salen)py	10.550	3.338	5.484	6.997	0.976

This splitting Δ_{xz} has been evaluated in the following way. From eq. 10, assuming $\Delta \gg \lambda$ and therefore $b \ll a \approx 1$ we have $\Delta g = g_z - g_y \approx 2c \approx \lambda/\Delta_{xz}$. Using the g values of Co(salen) (*75*), $\Delta_{xz} = 5\,000\ \text{cm}^{-1}$ is obtained.

In order to calculate state energies, interelectronic repulsion has to be included. This will be discussed later. The resulting ground states are $|yz, {}^2A_2\rangle$ for Co(salen), and $|z^2, {}^2A_1\rangle$ for $Co(salen)_2$ and Co(salen)py, in agreement with the experimental facts.

b) SCCC-MO Calculations

The SCCC-MO model is an extension of *Hoffmann's* original EHT (*39*), including charge iteration on the central metal atom (*21*). The coulomb parameters of the ligand atoms used in these calculations were taken from (*71*). The coulomb parameters of Co(II) are the SCCC values obtained through quadratic charge iterations using VOIP curves given by *Basch* et al. (*3*). "Single Zeta" STO's (*19*) were used for the li-

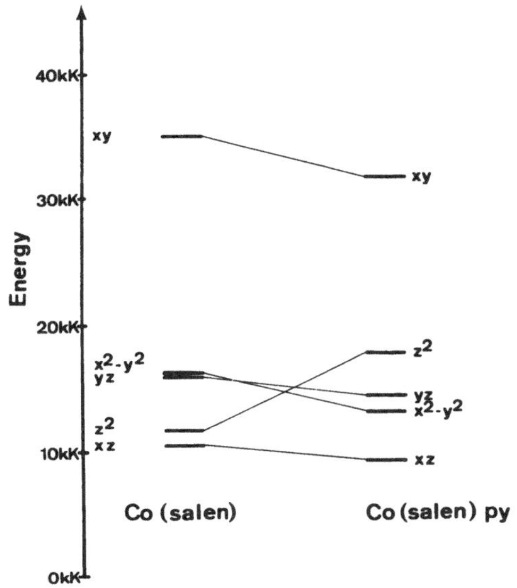

Fig. 10. d orbital energies for Co(salen) and Co(salen)py

gand orbitals and a "Double Zeta" expression for the Co(II) 3d orbitals. Metal 4 s and 4 p orbitals were represented by orthogonalized „Multiple Zeta" STO's (*64*).

Since the departure from C_{2v} symmetry is small, as shown by X-ray analysis (*65*), this idealized geometry has been used. The bond lengths and bond angles were chosen as follows:

R(Co–N)	= 1.90 Å	R(N–C)	= 1.30 Å
R(Co–O)	= 1.92 Å	R(O–C)	= 1.32 Å
α(N–Co–O)	= 93.7°	R(C–C)	= 1.49 Å

The base in five-fold coordinated complexes is represented by NH_3 (R : Co–N = 2.10 Å). The results of the calculation for Co(acacen) are displayed in Figures 11 and 12.

Four of the five d orbitals interact weakly with the ligand system and remain relatively close to the d orbital coulomb energy (– 12.26 eV). (1) d_{xz} overlaps very weakly with a b_1 π orbital of the ligand. (2) d_{yz} interacts strongly with an a_2 π orbital of the ligand with matching energy. It is this different interaction which explains the calculated orthorhombic splitting of about 0.2 eV in agreement with the observed orthorhombic magnetic tensors. (3) $d_{x^2-y^2}$ overlaps weakly with remote carbon atoms of the acacen ring. (4) d_{z^2} requires special consideration. Without an axial base the d_{z^2} orbital interacts mainly *in-plane* with the a_1 σ orbital of the ligand, as shown in Figure 11. When an axial ligand is present, in addition an *axial* σ interaction with the p_z orbital of the base destabilizes the d_{z^2} metal orbital strongly (Figs. 13 and 14).

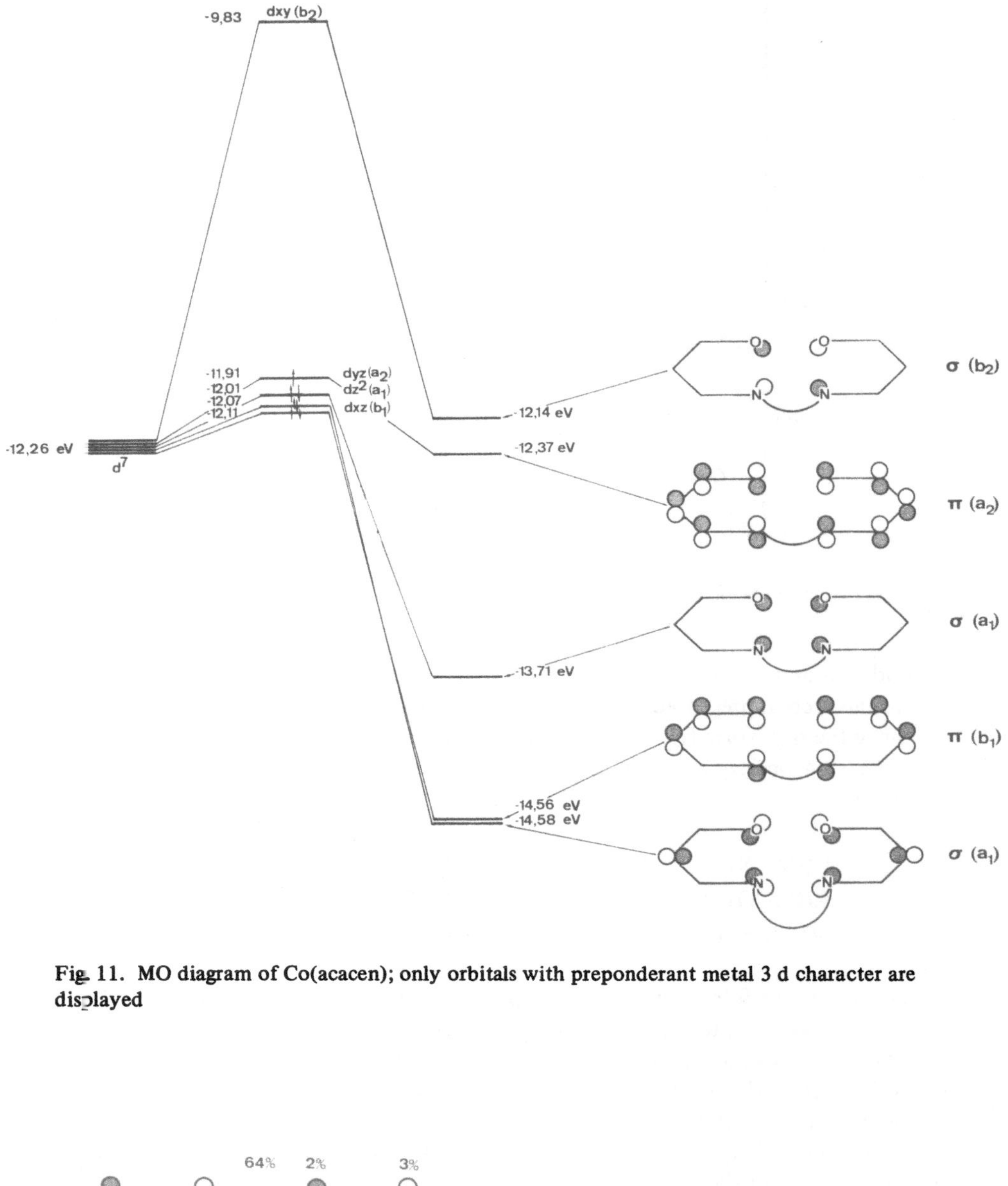

Fig. 11. MO diagram of Co(acacen); only orbitals with preponderant metal 3 d character are displayed

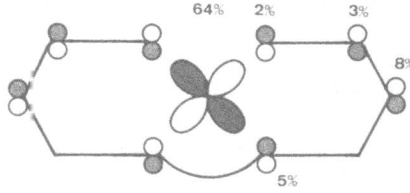

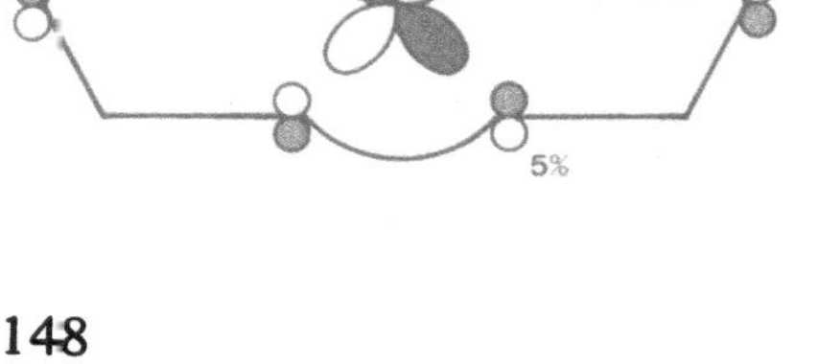

Fig. 12. HOMO of Co(acacen)

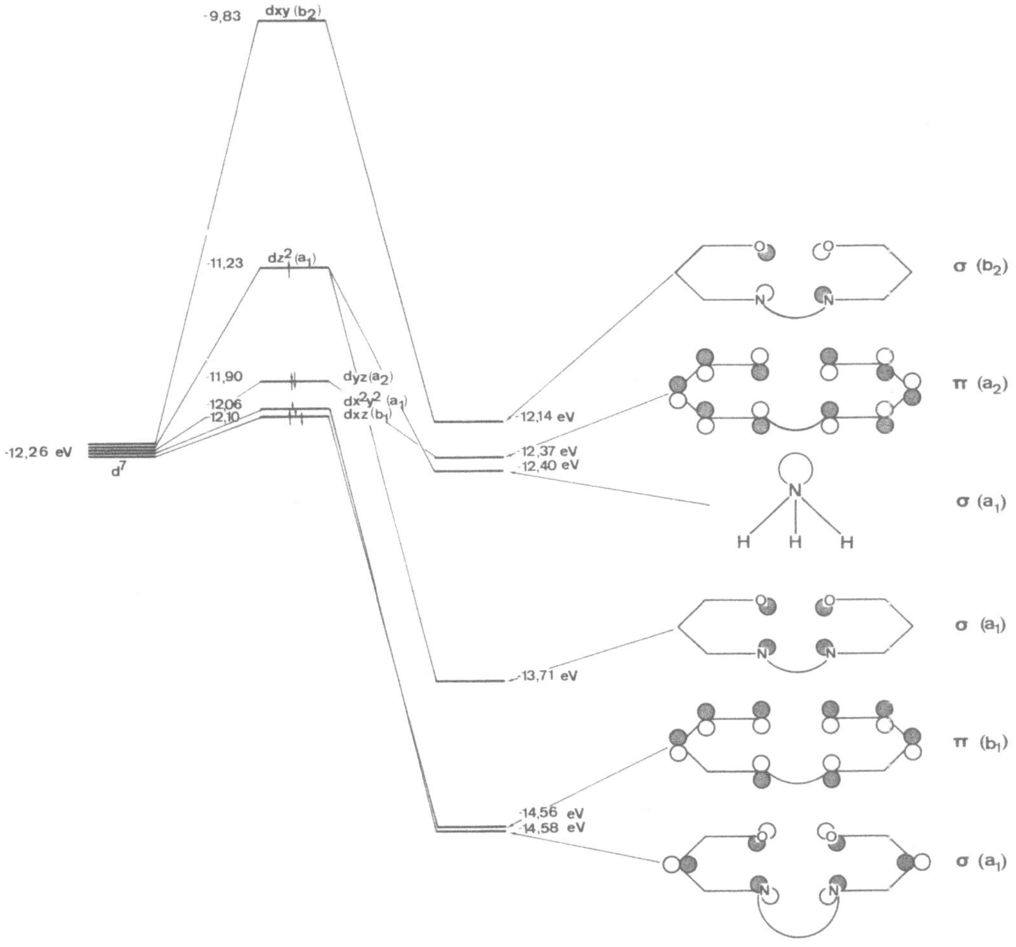

Fig. 13. MO diagram of Co(acacen)NH_3; only orbitals with preponderant metal 3d character are displayed

It is this latter interaction which is responsible for the change of the ground state upon coördination of the axial base.

The d_{xy} orbital, which interacts strongly with the b_2 σ orbital of the ligand, lies about 2 eV above the set of the other four d orbitals. Despite this large energy separation, the d_{xy} orbital plays an important role when state energies are considered. Promotion of an electron from a lower lying doubly occupied orbital creates a quartet state for which the spin-pairing energy approximately balances the promotion energy. These low lying quartet states play an important role in the interpretation of the A tensor.

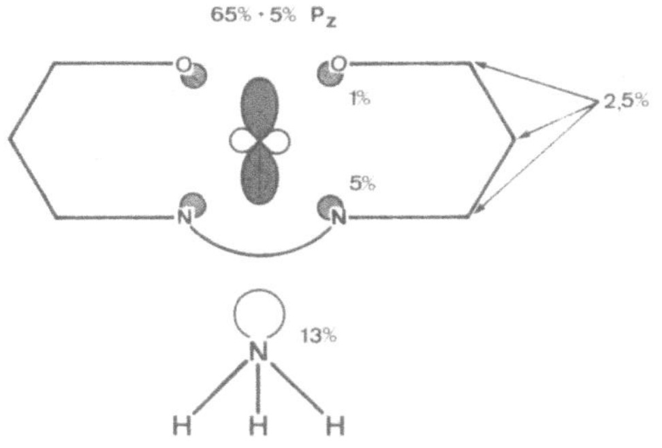

Fig. 14. HOMO of $Co(acacen)NH_3$

c) INDO-UHF Calculations (28)

A calculation for Co(acacen) and $Co(acacen)NH_3$, using the INDO-UHF (*63*) method, was carried out by *Fantucci* and *Valenti*. This approximation has the advantage over other similar methods of readily permitting the evaluation of spin densities. However, the UHF method gives eigenfunctions which are linear combinations of different spin states. In order to construct eigenfunctions with eigenvalues of S^2 not too far from the pure spin state value, a *single annihilation operator* has to be applied (*1*). The calculated spin density distribution of Co(acacen) and $Co(acacen)NH_3$ is taken from (*28*) (Tables 4 and 5).

The analysis of the spin density distribution shows that the unpaired electron is mainly localized in the d_{yz} orbital of the cobalt atom is Co(acacen), and in the d_{z^2} orbital in $Co(acacen)NH_3$. This conclusion is in complete agreement with the SCCC-MO calculation, and with the interpretation of the EPR and NMR spectra.

d) Ab initio LCAO-MO SCF Calculations (23)

Ab initio calculations of Co(acacen) and Co(acacen)L (L = H_2O, CO, CN and imidazole) have been reported, using a minimal *Gaussian* basis set. The calculations have been carried out in order to determine the ground state of Co(acacen). From SCF calculations for different electronic configurations, $|z^2, {}^2A_1\rangle$ emerges as the ground state (Fig. 15). This ground state cannot explain the sequence of the g components $g_x > g_z > g_y$ obtained from unambiguously assigned single crystal EPR measurements (cf. Chapter III), which shows that even the results of *ab initio* calculations have to be consumed *cum grano salis*.

When an axial ligand is present, the d_{z^2} orbital is destabilized and, hence, the $|z^2\ {}^2A_1\rangle$ state energy is lowered in agreement with experimental results.

Table 4. Spin density distribution in Co(acacen) (*28*)

Atom[a]	Orbitals								
	s	x	y	z	z^2	xz	yz	$x^2 - y^2$	xy
Co	0.0016	0.0008	0.0024	0.0045	0.0006	0.0042	0.9683	0.004	0.0234
O	− 0.0008	− 0.0042	− 0.0007	0.0076					
N	− 0.0018	− 0.0036	− 0.0020	0.0005					
C_1	− 0.0002	0.0000	− 0.0002	− 0.0022					
C_2	0.0003	0.0001	0.0001	0.0049					
C_3	− 0.0002	0.0001	− 0.0001	0.0005					

[a] C_1 is connected with O, C_3 with N and C_2 with C_1 and C_2.

Table 5. Spin density distribution in Co(acacen)NH_3 (*28*)

Atom[a]	Orbitals								
	s	x	y	z	z^2	xz	yz	$x^2 - y^2$	xy
Co	0.0217	0.0014	0.0014	0.0102	0.9254	0.0028	0.0002	0.0005	0.0231
O	− 0.0009	− 0.0010	− 0.0003	0.0020					
N_1	− 0.0014	− 0.0021	− 0.0012	− 0.0002					
C_1	0.0003	0.0000	0.0000	− 0.0009					
C_2	0.0000	0.0000	0.0001	0.0006					
C_3	0.0004	0.0001	0.0002	0.0000					
N_2[b]	0.0132	0.0007	0.0007	− 0.0002					

[a] C_1 is connected with O, C_3 with N_2 and C_2 with C_1 and C_2.
[b] N_2 is nitrogen atom of the axial base.

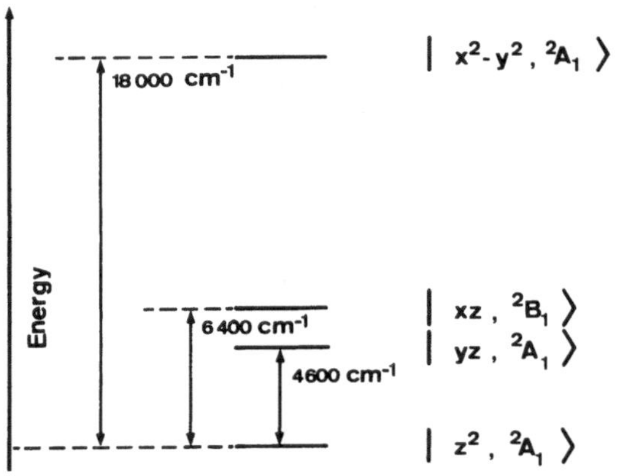

Fig. 15. SCF configuration energies of Co(acacen) (*23*)

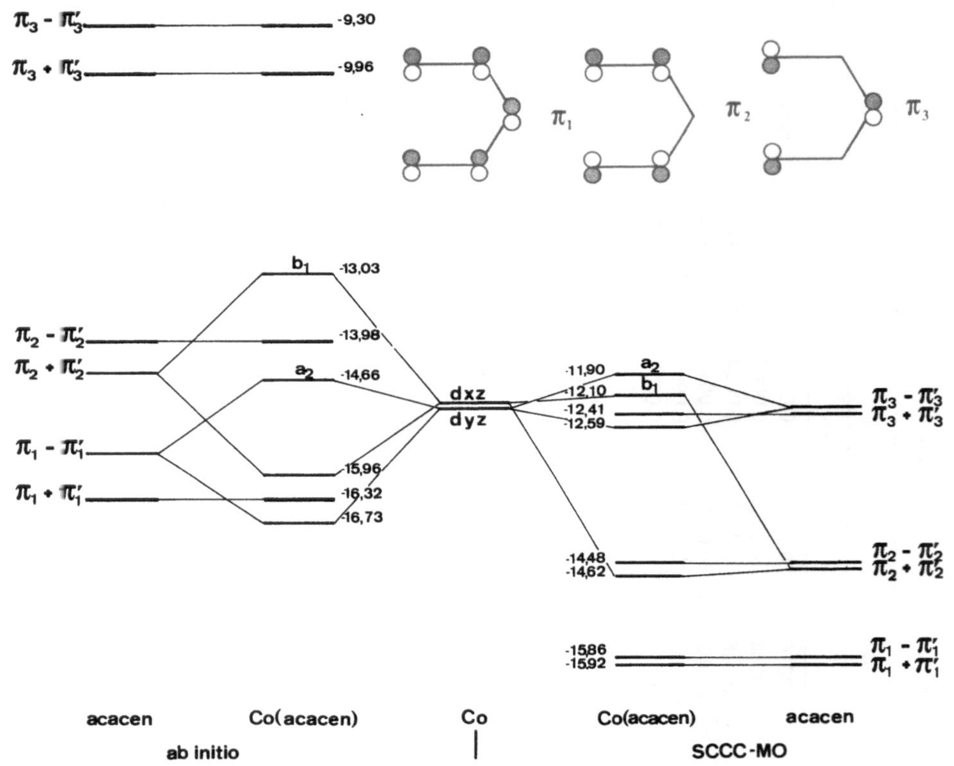

Fig. 16. Comparison of the orthorhombic splittings obtained from ab initio and SCCC-MO calculations

A further difference between the results of the *ab initio* calculation and those of the semi-empirical methods concerns the interaction between the metal d_{xz} and d_{yz} orbitals, and the corresponding ligand π orbitals. The *ab initio* calculation puts the a_2 (d_{yz}) orbital below the b_1 (d_{xz}) orbital, which is the opposite sequence to that obtained by the semi-empirical methods. The reason for this difference is that the *ab initio* energies of the 3d orbitals are between those of the π_1 and π_2 combinations, whereas the energies of the π_3 combinations are about 5 eV higher (Fig. 16). In the case of the SCCC-MO calculations, the 3d orbital energies almost match the energies of the π_3 combination. The resulting π interaction with the metal d_{yz} orbital is consequently much stronger than that with the metal d_{xz} orbital. Hence, the a_2 (d_{yz}) orbital is placed above the b_1 (d_{xz}) orbital (Fig. 16).

2. *State Energies*

Having established the ligand-field splitting, the calculation of the state energies of Co complexes is quite straightforward. It can be done by two different methods, either in a strong field scheme using published matrix elements (*60, 9, 61*), or in a weak field scheme using irreducible tensorial sets as described by *Harnung* and *Schäfer* (*35*).

Using the AOM bonding parameters of Table 3 and the electrostatic matrix elements for d^3 (reversing the sign of all ligand field parameters), *Hitchman* calculated the state energies of Co(salen), $[Co(salen)]_2$ and Co(salen)py (*37, 38*). The Racah parameters $B = 750\ cm^{-1}$ and $C = 3\,150\ cm^{-1}$ used in this calculation have been reduced compared to the free ion values to account for the decrease in effective nuclear charge on the metal, and for electron delocalization into the ligand orbitals. Results of these calculations are given in Figure 17. Following a weak field approach, a computer program CRISTF, based on irreducible tensorial sets, was developed (*22*). Using the SCCC-MO orbital energies, the calculations yield results which are in agreement with those obtained in the strong field approximation.

3. *Interpretation of the EPR Parameters*

An extensive account of the theory of the Spin Hamiltonian parameters for low spin Co(II) complexes has been given by *McGarvey* (*50*). This approach is based on third order perturbation theory, involving all excited states giving a spin-orbit contribution to the $|az^2 + b(x^2 - y^2), {}^2A_1\rangle$, or the $|yz, {}^2A_2\rangle$ ground state respectively. The formulae obtained for the $\tilde{g}$ and $\tilde{A}$ tensors in case of non axial symmetry depend upon a large number (12) of adjustable parameters, and a "judicious choice of assumptions" is required to make use of these equations. A simpler approach is therefore indicated.

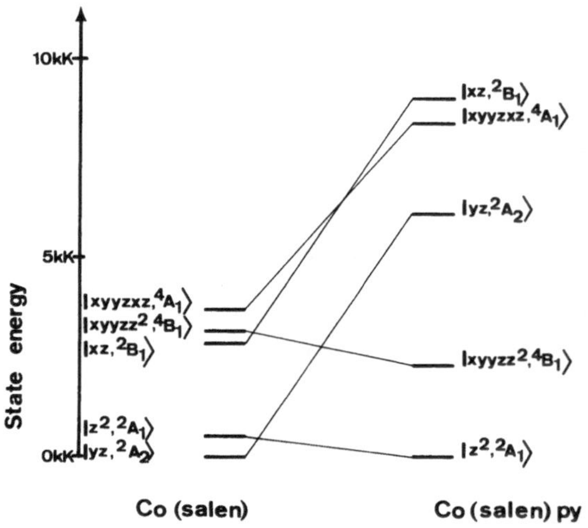

Fig. 17. State energies of Co(salen) and Co(salen)py, only states relevant to the interpretation of EPR parameters are displayed *(37)*

As discussed in Chapter III, the main features of the experimental g and A values are already expressed in the simple "crossover model". In order to refine this model, the perturbation of the ground state wave function by further low lying states has to be considered. The perturbed ground state wave function can be represented as a linear combination of state functions interacting via spin-orbit coupling. The matrix elements of the *Zeeman* and *Hyperfine* operators can be easily interpreted in terms of the $\tilde{g}$ and $\tilde{A}$ tensors *(66)*. The expressions obtained show that g values depend mainly upon the expansion coefficients of the low lying $|yz, {}^2A_2\rangle$, $|z^2, {}^2A_1\rangle$ and $|xz, {}^2B_1\rangle$ states. The three state crossover model including the perturbation of $|xz, {}^2B_1\rangle$ therefore satisfactorily explains the experimental g values of four- and five-fold coordinated complexes *(32)*. From the expressions for the A values one can see that additional low lying quartet states have to be considered, i.e. $|yz\,xz\,xy, {}^4A_1\rangle$, which contribute significantly in case of four-fold coordinated complexes with a dominantly $|yz, {}^2A_2\rangle$ ground state, $|xy\,yz\,z^2, {}^4B_1\rangle$ in five-fold coordinated complexes with a dominantly $|z^2, {}^2A_1\rangle$ ground state.

Consequently, the new basis set which we consider for the elaboration of the *refined model* is:

$$\overset{\pm}{\phi}_1 := |E^{(1)}_{1/2} \pm \tfrac{1}{2}\rangle = a'|yz, {}^2A_2 \pm \tfrac{1}{2}\rangle + ib'|z^2, {}^2A_1 \mp \tfrac{1}{2}\rangle$$

$$\overset{\pm}{\phi}_2 := |E^{(2)}_{1/2} \pm \tfrac{1}{2}\rangle = b'|yz, {}^2A_2 \pm \tfrac{1}{2}\rangle - ia'|z^2, {}^2A_1 \mp \tfrac{1}{2}\rangle$$

$$\overset{\pm}{\phi}_3 := \mp i|xz, {}^2B_1 \pm \tfrac{1}{2}\rangle \qquad (5)$$

$$\overset{\pm}{\phi}_4 := \pm\frac{i}{2}|xy\,yz\,z^2, {}^4B_1 \mp \tfrac{3}{2}\rangle \mp \frac{i\sqrt{3}}{2}|xy\,yz\,z^2, {}^4B_1, \pm\tfrac{1}{2}\rangle$$

$$\overset{\pm}{\phi}_5 := \frac{i}{2}|xy\,yz\,xz, {}^4A_1 \pm\tfrac{3}{2}\rangle + \frac{i\sqrt{3}}{2}|xy\,yz\,xz, {}^4A_1 \mp\tfrac{1}{2}\rangle$$

where $a' = \cos\theta$, $b' = \sin\theta$ and $\theta = -\frac{1}{2}\tan^{-1}\frac{\sqrt{3}\,\lambda}{\Delta}(0 \leqslant \theta \leqslant \frac{\pi}{2})$. Δ is the energy separation between $|yz, {}^2A_2\rangle$ and $|z^2, {}^2A_1\rangle$, and λ the spin-orbit coupling constant. The perturbed ground state wave function is then:

$$|\pm\rangle = a\overset{\pm}{\phi}_1 + b\overset{\pm}{\phi}_2 + c\overset{\pm}{\phi}_3 + d\overset{\pm}{\phi}_4 + e\overset{\pm}{\phi}_5 \tag{6}$$

In order to determine the expansion coefficients, the following matrix elements have to be evaluated.

$$\begin{aligned}
&\langle\phi_2|\lambda\mathbf{L}\cdot\mathbf{S}|\phi_1\rangle = 0\\
&\langle\phi_3|\lambda\mathbf{L}\cdot\mathbf{S}|\phi_1\rangle = \frac{-a' + b'\sqrt{3}}{2}\lambda\\
&\langle\phi_4|\lambda\mathbf{L}\cdot\mathbf{S}|\phi_1\rangle = -\frac{b'}{2}\lambda\\
&\langle\phi_5|\lambda\mathbf{L}\cdot\mathbf{S}|\phi_1\rangle = -\frac{a'}{2}\lambda
\end{aligned} \tag{7}$$

The state energies relative to the energy of $|yz, {}^2A_2\rangle$ are

$$\begin{aligned}
&E(\phi_1) = \Delta\sin^2\theta + \frac{\sqrt{3}}{2}\lambda\sin 2\theta\\
&E(\phi_2) = \Delta\cos^2\theta - \frac{\sqrt{3}}{2}\lambda\sin 2\theta\\
&E(\phi_3) = \Delta_{xz}\\
&E(\phi_4) = \Delta + \Delta_{xy} - 4B - 4C = \Delta + Q\\
&E(\phi_5) = \Delta_{xy} - \Delta_{xz} - 12B - 4C = R
\end{aligned} \tag{8}$$

Δ_{xz} and Δ_{xy} are the energy separations, $d_{xz} - d_{yz}$ and $d_{xy} - d_{yz}$ respectively, B and C are Racah parameters. A representation of these energies vs. Δ is given in Figure 19. Using second-order perturbation theory the following coefficients are obtained.

$$a = N\cos\theta \approx \cos\theta$$

$$b = N\sin\theta \approx \sin\theta$$

$$c = -N\frac{\lambda(\sin\theta\sqrt{3} - \cos\theta)}{2(\Delta_{xz} - \Delta\sin^2\theta - \frac{\sqrt{3}}{2}\lambda\sin 2\theta} \approx \frac{\lambda(\sqrt{3}\sin\theta - \cos\theta)}{2(\Delta\sin^2\theta - \Delta_{xz})}$$

$$d = N\frac{\lambda\sin\theta}{2(\Delta + Q - \Delta\sin^2\theta - \frac{\sqrt{3}}{2}\lambda\sin 2\theta} \approx \frac{\lambda\sin\theta}{2(Q + \Delta\cos^2\theta)} \qquad (9)$$

$$e = N\frac{\lambda\cos\theta}{2(R - \Delta\sin^2\theta - \frac{\sqrt{3}}{2}\lambda\sin 2\theta} \approx \frac{\lambda\cos\theta}{2(R - \Delta\sin^2\theta)}$$

N = Normalization factor.

The new ground state wave function $|\pm\rangle$ yields the following expressions for the $\tilde{g}$ and $\tilde{A}$ components:

$$g_x = 2(a^2 + b^2 - c^2 + \tfrac{8}{3}d^2 + \tfrac{4}{3}e^2) + 4\sqrt{3}\,ab = 2T_x + 4\sqrt{3}\,ab$$

$$g_y = 2(a^2 - b^2 - c^2 - \tfrac{4}{3}d^2 + \tfrac{8}{3}e^2) + 4\sqrt{3}\,bc - \frac{8\sqrt{3}}{3}de = 2T_y + 4\sqrt{3}\,bc - \frac{8\sqrt{3}}{3}de$$

$$g_z = 2(a^2 - b^2 + c^2 - \tfrac{8}{3}d^2 + \tfrac{8}{3}e^2) + 4ac = 2T_z + 4ac$$

$$A_x = 2P(-\tfrac{2}{7}a^2 - \tfrac{1}{7}b^2 - \tfrac{1}{7}c^2 - \tfrac{16}{63}d^2 + 2\sqrt{3}\,ab + \tfrac{3}{7}ac + \frac{8}{7\sqrt{3}}ad - \frac{\sqrt{3}}{7}bc + \tfrac{2}{7}bd - \tfrac{2}{7}ce + \frac{10}{21\sqrt{3}}de - \frac{\kappa}{2}T_x)$$

$$A_y = 2P(\tfrac{1}{7}a^2 + \tfrac{1}{7}b^2 + \tfrac{2}{7}c^2 - \tfrac{4}{63}d^2 - \frac{\sqrt{3}}{7}ab - \tfrac{3}{7}ac + \frac{4}{7\sqrt{3}}ad + \tfrac{2}{7}ae + 2\sqrt{3}\,bc - \frac{4}{\sqrt{3}}de - \frac{\kappa}{2}T_y) \qquad (10)$$

$$A_z = 2P(\tfrac{1}{7}a^2 - \tfrac{2}{7}b^2 + \tfrac{1}{7}c^2 - \tfrac{8}{63}d^2 + \frac{\sqrt{3}}{7}ab + 2ac - \tfrac{2}{7}ae - \frac{\sqrt{3}}{7}bc + \tfrac{2}{7}bd + \tfrac{4}{7}ce - \frac{2}{21\sqrt{3}}de - \frac{\kappa}{2}T_z)$$

Within this model, a unified and quantitative explanation of the EPR parameters of four- and five-fold coordinated complexes is possible (Fig. 18).

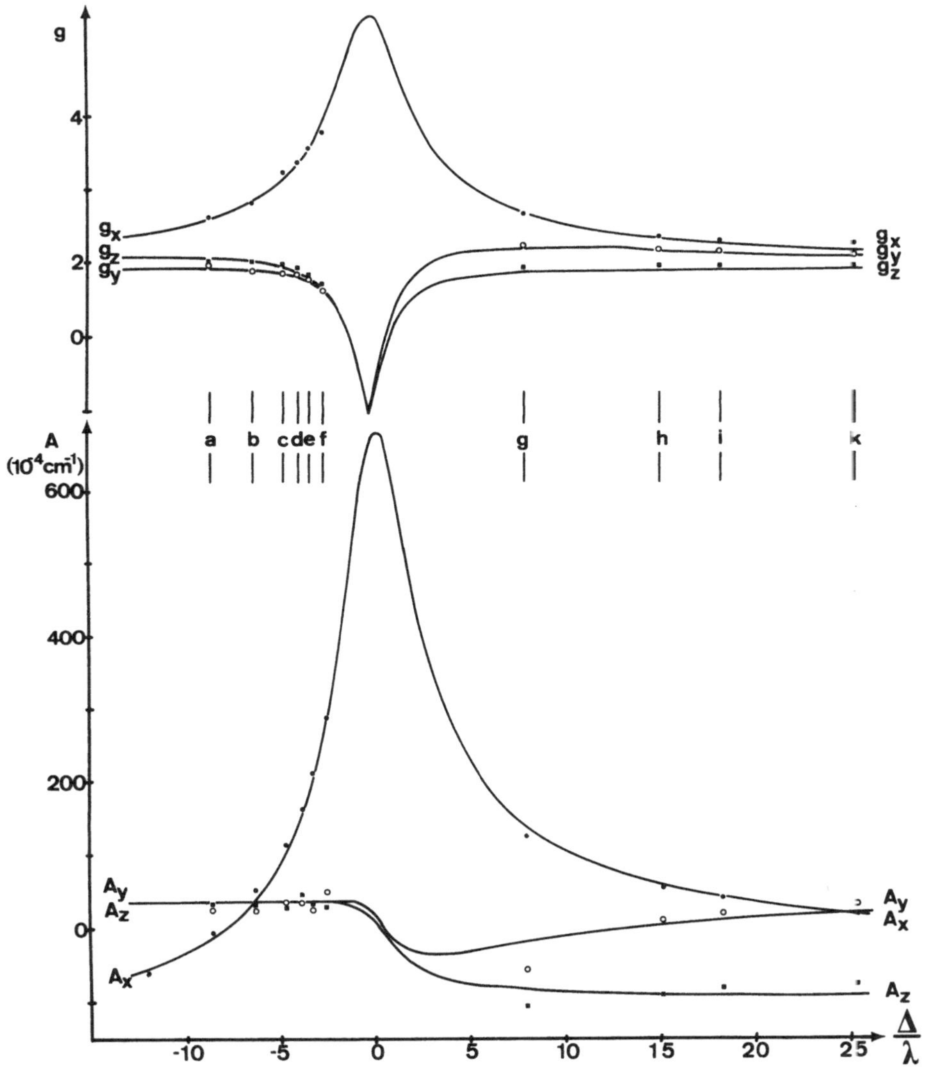

Fig. 18. EPR parameters in the refined model including quartet states as a function of Δ/λ. The solid lines represent the functions calculated from Eq. (10). The energy parameters were obtained through least square fitting of the EPR data of ten complexes: a) Co(amben); b) Co(tacacen); c) Co(acacen); d) Co(bzacacen); e) Co(CF_3acacen); f) Co(salen); g) $[Co(salen)]_2$; h) Co(bzacacen)py; i) Co(salen)$P(C_6H_5)_3$; k) Co(saphen)py_2. $P = 220.8 \cdot 10^{-4}$ cm^{-1}; $\kappa = 0.2$ (scaled with a^2); $\Delta_{xz} = 9.4\ \lambda$, $Q = 2.5\ \lambda$; $R = 2.2\ \lambda$

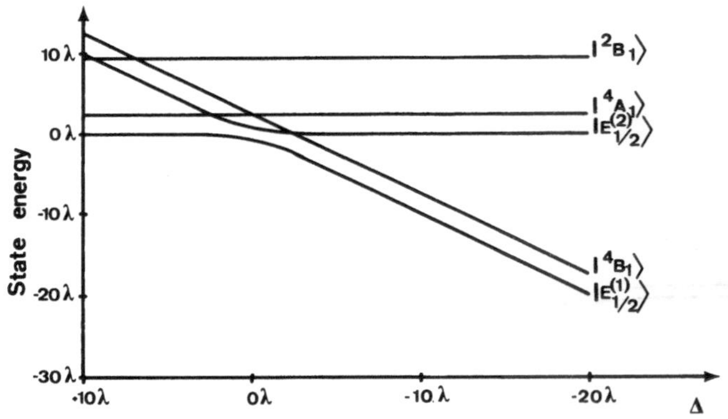

Fig. 19. State energies obtained from EPR measurements using the refined model

It should be noted that only the relative energy of d_{z^2} has to be varied for different complexes (Fig. 19). In most complexes, the important contributions to the EPR parameters arive from mixing of the ground state with excited states lying below 5 000 cm^{-1}.

VIII. Conclusions

In the present class of complexes of Co(II) with tetradentate ligands, which impose an approximately planar ligand field having nearly C_{2v} symmetry, the electronic states show some very peculiar properties.

Owing to the planar coordination, the set of five degenerate 3d orbitals of the free metal ion is split into a group of four MO's which lie close together and one which is distinctly higher in energy. This arrangement, which is undoubtedly also characteristic of complexes of other metal ions with similar ligands, gives rise to quite "normal" electronic states in the case of e.g. Cu(II) ($3d^9$) or Ni(II) ($3d^8$), where in the ground state the closely spaced group of four orbitals is completely filled. In the case of Co(II) (d^7), this group accommodates only 7 electrons, leaving a "one hole" configuration of the set. Due to the proximity of these orbitals, four doublet states of similar energies are expected. One of these states (the $|x^2 - y^2, {}^2A_1\rangle$ hole configuration) is placed appreciably higher by electron repulsion, leaving three doublets, namely $|yz, {}^2A_2\rangle$, $|z^2, {}^2A_1\rangle$ and $|xz, {}^2B_1\rangle$, close together. All these three states interact strongly by spin-orbit coupling since their energy separation is comparable to the spin-orbit coupling constant. This has dramatic effects on the EPR spectra, which show g values very far from 2.0023.

Not only are doublets very close to the ground state but also several quartets arising by promotion of one electron from the set of four orbitals to the higher lying level have similar energies because the spin pairing energy almost compensates the promotional energy. Therefore the unusual feature of these complexes is the fact that, within a range of about 5 000 cm^{-1} above the ground state doublet, two other doublets and four quartets are located. Hence small changes of the state energies due to small structural perturbations lead to significant variations of the EPR parameters.

A detailed analysis shows that the EPR spectra (and the NMR spectra) can be interpreted to a very good approximation by an energy diagram having two doublets and one quartet within 2000 cm^{-1} above the predominantly $|yz, {}^2A_2\rangle$ ground state. Higher lying states have almost no influence on the EPR parameters.

Base adducts, i.e. five-fold coordinated species, show very different EPR spectra because another doublet state, $|z^2, {}^2A_1\rangle$, becomes the ground state. The doublets are now somewhat more widely spaced, but still rather close together, and some quartets are even nearer to the ground state than in the four coordinated complexes.

Electronic spectra observed in the four- and five-fold coordinated complexes concern states quite different from those of EPR spectra. Since electronic spectra of single crystals at low temperatures are lacking, no completely definite assignment of absorption bands can be given. An interpretation of the electronic spectrum of a particular complex in the series which can claim some reasonable degree of probability of being correct is only possible if the energies of the higher lying states are explicable in the same theoretical model as a completely assigned EPR spectrum of the same complex.

This class of four- and five-fold coordinated complexes of Co(II) exhibits the unusual property of having a number of electronically excited states at very low energies with the same, and with higher, spin-multiplicities as the ground state. Further investigations of this type of complex should be carried out, taking into account this peculiar electronic structure.

IX. Appendix

Spectroscopic and magnetic properties of low spin Co(II) complexes with *Schiff* bases and related ligands.

Abreviations:

s.c.: single crystal, p.c.: polycrystalline sample, g.: glass, n.p.: nematic phase, sh.: shoulder, refl.: reflectance spectrum

py: pyridine, N-meim: N-methylimidazole, dmso: dimethylsulfoxide, dmf: dimethylformamide, thf: tetrahydrofurane

---

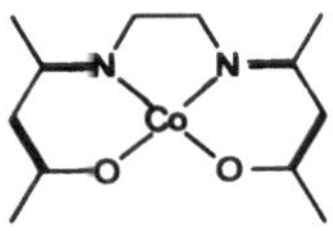

Co(acacen)

EPR: Ni(acacen) subl., 113 K, s.c. *(12)*

g_x[a]) 3.26 A_x[a]) $115.8 \cdot 10^{-4}$ cm^{-1}
g_y[a]) 1.88 A_y[a]) $37.5 \cdot 10^{-4}$ cm^{-1}
g_z 2.00 A_z $34.5 \cdot 10^{-4}$ cm^{-1}

[a]) axes of the g- and A-tensor rotated 45° in the molecular plane

Ni(acacen) subl., 110 K, p.c. *(66)*

g_1 3.190 (5) A_1 115 (1) $\cdot 10^{-4}$ cm^{-1}
g_2 2.010 (5) A_2 37 (1) $\cdot 10^{-4}$ cm^{-1}
g_3 1.925 (5) A_3 31 (1) $\cdot 10^{-4}$ cm^{-1}

Ni(acacen) · 1/2 H_2O, 110 K, p.c. *(66)*

g_1 3.228 (5) A_1 128.5 (2) $\cdot 10^{-4}$ cm^{-1}
g_2 2.003 (5) A_2 40 (3) $\cdot 10^{-4}$ cm^{-1}
g_3 1.920 (5) A_3 33 (1) $\cdot 10^{-4}$ cm^{-1}

Pd(acacen) subl., 110 K, p.c. *(66)*

g_1 3.154 (5) A_1 111 (1) $\cdot 10^{-4}$ cm^{-1}
g_2 2.03 (1) A_2 34 (2) $\cdot 10^{-4}$ cm^{-1}
g_3 1.928 (5) A_3 29 (2) $\cdot 10^{-4}$ cm^{-1}

Pd(acacen) · 1/2 H_2O, 110 K, p.c. *(66)*

g_1 3.191 (5) A_1 117 (2) $\cdot 10^{-4}$ cm^{-1}
g_2 2.02 (1) A_2 37 (2) $\cdot 10^{-4}$ cm^{-1}
g_3 1.920 (8) A_3 31 (2) $\cdot 10^{-4}$ cm^{-1}

Phase 7 A[b]), 110 K, n.p. *(76)*

g_x 3.16 (1) A_x 111 (7) $\cdot 10^{-4}$ cm^{-1}
g_y 1.92 (2) A_y –
g_z 1.99 (2) A_z –

[b]) Nematic phases licristal, *Merck*, Darmstadt

N-(4'methoxybenzylidene)-4-butylaniline, 77 K, n.p. *(40)*

g_x 2.928 A_x $100.3 \cdot 10^{-4}$ cm^{-1}
g_y 1.934 A_y $32.5 \cdot 10^{-4}$ cm^{-1}
g_z 2.01 A_z $32.8 \cdot 10^{-4}$ cm^{-1}

Magn. Susc.: 100–300 K, s.c. *(56)*

K_x 3330 · 10^{-6} cm^{-3} mol^{-1} (300 K)
K_y 1790 · 10^{-6} cm^{-3} mol^{-1} (300 K)
K_z 1330 · 10^{-6} cm^{-3} mol^{-1} (300 K)

NMR: dmso-d_6, 280–400 K *(69)*

Absorption Spectrum: $CHCl_3$, 3.5–15 kK *(66)*

3.940 (71), 4.100 sh, 5.250 sh, 8.400 (13) kK (ϵ)

P.E.: Ionisation energies < 13 eV *(27)*

8.23, 9.35, 9.57, 9.81, 10.73,
11.01, 11.05, 11.81, 12.21, 12.89 eV

Co(acacen)py

EPR: py, 100 K, g. *(6)*

g_1 2.51 A_1 68.4 · 10^{-4} cm^{-1}
g_2 2.26 A_2 12 · 10^{-4} cm^{-1}
g_3 2.01 A_3 94 · 10^{-4} cm^{-1}

1% py-toluene, 77 K, g. *(10)*

g_1 2.449 A_1 47.3 Gauss
g_2 2.236 A_2 –
g_3 2.013 A_3 100 Gauss A^N 15.0 Gauss

N-(4'methoxybenzylidene)-4-butylaniline, 77 K, n.p. *(40)*

g_x 2.435 A_x 56.2 · 10^{-4} cm^{-1}
g_y 2.225 A_y 12.5 · 10^{-4} cm^{-1}
g_z 2.012 A_z 92.3 · 10^{-4} cm^{-1} A^N 14.7 · 10^{-4} cm^{-1}

Absorption spectrum: py, 4–16 kK *(6)*

5.500, 8.400, 13.300 kK

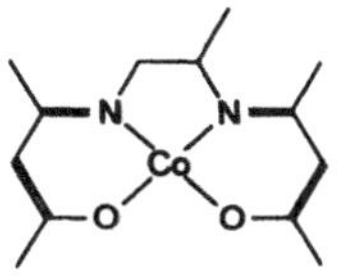

Co(acacpn)

EPR: toluene, 100 K, g. *(70)*

g_x 3.1619 A_x 75.5 Gauss
g_y 1.8968 A_y 33.4 Gauss
g_z 2.0117 A_z 37.6 Gauss

Absorption spectrum: C_2Cl_4, 3.5–35 kK *(70)*

3.970 (96), 4.300 sh (57), 4.870 sh (23), 5.550 (16),
8.500 (20), 20.600 (360), 22.000 sh (1100),
26.800 (6500), 30.000 (6150), 34.000 (13000) kK (ϵ)

C.D.: Co(acac(–)pn), C_2Cl_4, 8.3–40 kK *(70)*

15.630 (+ 0.003), ~ 18.000 sh (– 0.33), ~ 19.500 (+),
20.700 (– 2.27), 22.400 (– 1.50), 25.450 (2.14),
28.400 (4.55), 32.250 (32.4), 35.700 (15.60) kK ($\Delta\epsilon$)

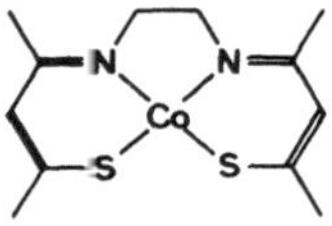

Co(tacacen)

EPR: Ni(tacacen), 110 K, p.c. (*66*)

g_1 3.047 (5) A_1 95 (1) $\cdot 10^{-4}$ cm^{-1}
g_2 1.98 (1) A_2 23.1 (5) $\cdot 10^{-4}$ cm^{-1}
g_3 1.945 (5) A_3 27.8 (5) $\cdot 10^{-4}$ cm^{-1}

Pd(tacacen), 110 K, p.c. (*66*)

g_1 3.170 (5) A_1 125 (2) $\cdot 10^{-4}$ cm^{-1}
g_2 1.93 (1) A_2 20 (2) $\cdot 10^{-4}$ cm^{-1}
g_3 1.910 (5) A_3 26 (2) $\cdot 10^{-4}$ cm^{-1}

Zn(tacacen), 110 K, p.c. (*66*)

g_1 2.848 (5) A_1 56 (1) $\cdot 10^{-4}$ cm^{-1}
g_2 2.026 (8) A_2 27.3 (5) $\cdot 10^{-4}$ cm^{-1}
g_3 1.910 (5) A_3 23.8 (5) $\cdot 10^{-4}$ cm^{-1}

Magn. Susc: 200–300 K, s.c. (*56*)

K_x 2776 $\cdot 10^{-6}$ cm^3 mol^{-1} (300 K)
K_y 2198 $\cdot 10^{-6}$ cm^3 mol^{-1} (300 K)
K_z 951 $\cdot 10^{-6}$ cm^3 mol^{-1} (300 K)

Absorption spectrum: $CHCl_3$, 3.5–15 kK (*66*)

4.200 (10), 5.130 (18), 10.000 (25) kK (ϵ)

Co(tacacen)py

EPR: 1% py-toluene, 77 K, g. (*10*)

g_1 2.443 A_1 –
g_2 2.169 A_2 –
g_3 2.003 A_3 77.6 Gauss A^N 15.0 Gauss

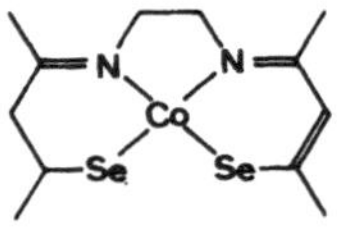

Co(seacacen)

EPR: Ni(seacacen), 110 K, p.c. (*66*)

g_1 3.330 (5) A_1 161.5 (3) $\cdot 10^{-4}$ cm^{-1}
g_2 1.930 (3) A_2 26.9 (1) $\cdot 10^{-4}$ cm^{-1}
g_3 1.925 (3) A_3 14.3 (1) $\cdot 10^{-4}$ cm^{-1}

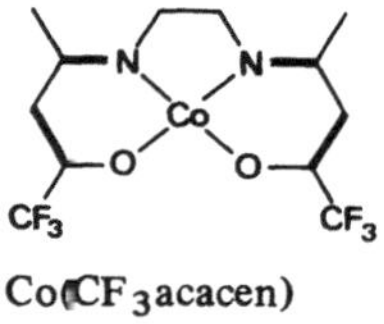

Co(CF$_3$acacen)

EPR: Ni(CF$_3$acacen), 100 K, s.c. (*44*)

g_x 3.562 (5) A_x 209.0 (1) $\cdot 10^{-4}$ cm^{-1}
g_y 1.80 (1) A_y 31.0 (5) $\cdot 10^{-4}$ cm^{-1}
g_z 1.86 (1) A_z 25.0 (5) $\cdot 10^{-4}$ cm^{-1}

Ni(CF$_3$acacen), 100 K, p.c. (*6*)

g_1 3.53 A_1 210.5 $\cdot 10^{-4}$ cm^{-1}
g_2 1.88 A_2 29.3 $\cdot 10^{-4}$ cm^{-1}
g_3 1.85 A_3 48.9 $\cdot 10^{-4}$ cm^{-1}

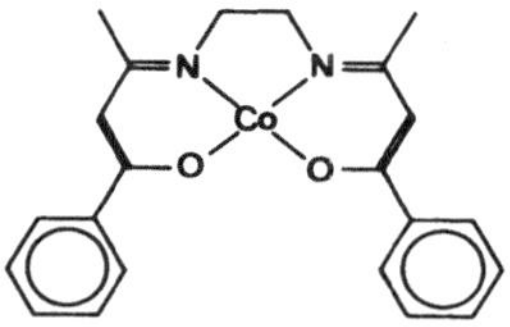

Co(bzacacen)

EPR: Ni(bzacacen), 100 K, s.c. (47)

g_x 3.372 (2) A_x 165.6 (2) · 10^{-4} cm^{-1}
g_y 1.882 (1) A_y 37 (5) · 10^{-4} cm^{-1}
g_z 1.954 (6) A_z 47 (5) · 10^{-4} cm^{-1}

N-(4'methoxybenzylidene)-4-butylaniline, 100 K, n.p. (40)

g_x 3.084 A_x 104.1 · 10^{-4} cm^{-1}
g_y 1.937 A_y 32.0 · 10^{-4} cm^{-1}
g_z 2.01 A_z 32.8 · 10^{-4} cm^{-1}

Co(bzacacen)py

EPR: 1% py-toluene, 77 K, g. (10)

g_1 2.444 A_1 45.0 Gauss
g_2 2.231 A_2 –
g_3 2.014 A_3 98.0 Gauss A^N 15.0 Gauss

N-(4'methoxybenzylidene)-4-butylaniline, 77 K, n.p. (40)

g_x 2.428 A_x 55.2 · 10^{-4} cm^{-1}
g_y 2.22 A_y 13.5 · 10^{-4} cm^{-1}
g_z 2.012 A_z 92.1 · 10^{-4} cm^{-1} A^N 14.9 · 10^{-4} cm^{-1}

Co(bzacacen)N-meim

EPR: 70% toluene, 20% dmf, 10% N-mein, 77 K, g. (10)

g_1 2.442 A_1 44 Gauss
g_2 2.257
g_3 2.013 A_3 98 Gauss A^N 16 Gauss

Co(bzacacen)$P(C_6H_5)_3$

EPR: 70% toluene, 20% dmf, 10% $P(C_6H_5)_3$ 77 K, g. (10)

g_1 2.319 A_1 –
g_2 2.250 A_2 –
g_3 1.958 A_3 89 Gauss

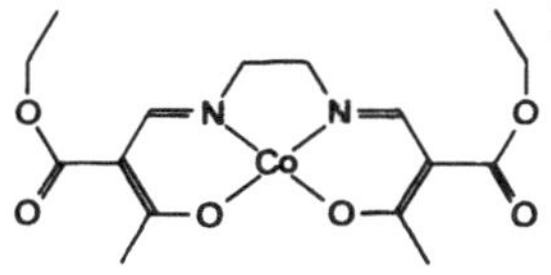

Co($a_2$en)

EPR: Ni($a_2$en), 110 K, p.c. (66)

g_1 3.907 (5) A_1 295 (2) · 10^{-4} cm^{-1}
g_2 1.775 (8) A_2 29 (2) · 10^{-4} cm^{-1}
g_3 1.69 (2) A_3 47 (2) · 10^{-4} cm^{-1}

Pd($a_2$en), 110 K, p.c. (66)

g_1 4.239 (5) A_1 393 (2) · 10^{-4} cm^{-1}
g_2 1.61 (2) A_2 –
g_3 1.58 (2) A_3 53 (8) · 10^{-4} cm^{-1}

Phase 5 A[c]), 110 K, n.p. (66)

g_1 3.74 (2) A_1 260 (10) · 10^{-4} cm^{-1}
g_2 1.85 (5) A_2 –
g_3 1.75 (5) A_3 –

[c]) Nematic phases licristal, *Merck*, Darmstadt

Absorption spectrum: $CHCl_3$, 3.5–15 kK (66)

4.000 (103), 4.400 sh, 4.880 sh, 8.000 (16) kK (ε)

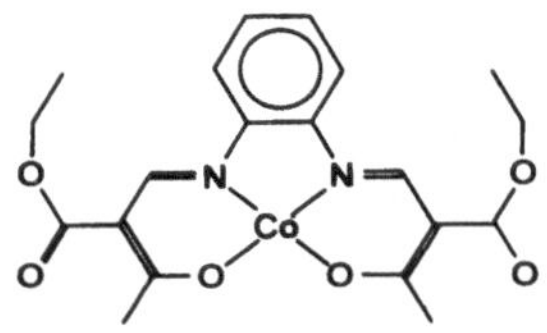

Co($a_2$phen)

EPR: Ni($a_2$phen), 110 K, p.c. (*66*)

g_1 5.147 (1) A_1 615 (1) $\cdot 10^{-4}$ cm^{-1}
g_2 0.7 (2) A_2 –
g_3 0.7 (2) A_3 –

Pd($a_2$phen), 110 K, p.c. (*66*)

g_1 4.546 (2) A_1 479.7 (5) $\cdot 10^{-4}$ cm^{-1}
g_2 1.33 (2) A_2 –
g_3 1.33 (2) A_3 80 (5) $\cdot 10^{-4}$ cm^{-1}

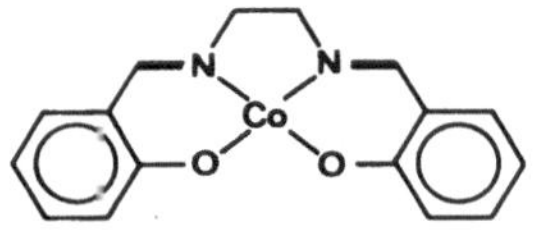

Co(salen)

EPR: Ni(salen), 130 K, s.c. (*75*)

g_x 3.805 (5) A_x 291 (1) $\cdot 10^{-4}$ cm^{-1}
g_y 1.66 (1) A_y 52 (26) $\cdot 10^{-4}$ cm^{-1}
g_z 1.74 (1) A_z 30 (5) $\cdot 10^{-4}$ cm^{-1}

Zn(salen)[d]), 100 K, p.c. (*7*)

g_1 2.69 A_1 125.9 $\cdot 10^{-4}$ cm^{-1}
g_2 2.31 A_2 58.4 $\cdot 10^{-4}$ cm^{-1}
g_3 2.002 A_3 110.4 $\cdot 10^{-4}$ cm^{-1}

[d]) Co(salen) inactiv, $[Co(salen)]_2$ (*37*)

$CHCl_3$, 100 K, g. (*7*)

g_1 3.20 A_1 133 $\cdot 10^{-4}$ cm^{-1}
g_2 2.14 A_2 90 $\cdot 10^{-4}$ cm^{-1}
g_3 1.91 A_3 22.3 $\cdot 10^{-4}$ cm^{-1}

$CHCl_3$, 77 K, g. (*15*)

g_1 3.34, 3.28, 3.23 A_1 153, 141, 125 $\cdot 10^{-4}$ cm^{-1}

Phase 5[e]), 110 K, n.p. (*76*)

g_x 3.32 (6) A_x 155 (20) $\cdot 10^{-4}$ cm^{-1}
g_y 1.84 (2) A_y –
g_z 1.93 (7) A_z –

[e]) Nematic phases licristal, *Merck*, Darmstadt

Pd(salen), 110 K, p.c. (*66*)

g_1 4.044 (5) A_1 351.4 (5) $\cdot 10^{-4}$ cm^{-1}
g_2 1.61 (2) A_2 –
g_3 1.58 (2) A_3 –

Magn. Susc.: Co(salen) active, 100–400 K, p.c. (*8*)

χ_m 2650 $\cdot 10^{-6}$ cm^3 mol^{-1} (296 K)
Co(salen) inactive, 100–400 K, p.c.
χ_m 2170 $\cdot 10^{-6}$ cm^3 mol^{-1} (290 K)

Co(salen) inactive, 88.5–351.5 K, p.c. (*25*)
χ_m 1992 $\cdot 10^{-6}$ cm^3 mol^{-1} (280 K)

$[Co(salen)]_2$, 10–300 K, p.c. (*11*)
$[Co(salen)]_2$, 100–320 K, s.c. (*56*)

K_x 2669 · 10^{-6} cm^3 mol^{-1} (320 K)
K_y 1813 · 10^{-6} cm^3 mol^{-1} (320 K)
K_z 1347 · 10^{-6} cm^3 mol^{-1} (320 K)

NMR: $CDCl_3$, 220–350 K *(53)*
$dmso-d_6$, 275–350 K *(69)*

Absorption spectrum:
Cosalen (active), refl., 300 K, 4–15 kK *(7)*
5.36, 5.80 sh, 8.40, 9.00 sh, 10.00 sh, 14.50 kK

Cosalen (inactive), refl., 300 K, 4–15 kK *(7)*
6.00, 7.50, 11.80, 12.80 kK

Co(salen), $CHCl_3$, 300 K, 3.5–25 kK *(37)*
3.90 (65), 8.30 (16) kK (ϵ)

$[Co(salen)]_2$, refl., 77 K, 4–25 kK *(37)*
6.10, 11.50 kK

P.E.: Co 2 $p_{1/2}$ 796.0 (1), 794.8 (2) eV *(4)*
Co 2 $p_{3/2}$ 779.9 (1) eV
Co 2 $p_{1/2,3/2}$ 60.1 (2) eV
N $ls_{1/2}$ 398.9 (2) eV
O $ls_{1/2}$ 530.9 (2) eV

Co(salen)py

EPR: py, 77 K, g. *(59)*
g_1 2.354 $A_1 < 20 \cdot 10^{-4}$ cm^{-1}
g_2 2.27 $A_2 < 20 \cdot 10^{-4}$ cm^{-1}
g_3 2.028 A_3 $77 \cdot 10^{-4}$ cm^{-1}

Zn(salen)py, 110 K, p.c. *(7)*
g_1 2.41 A_1 $41.4 \cdot 10^{-4}$ cm^{-1}
g_2 2.24 A_2 $24.4 \cdot 10^{-4}$ cm^{-1}
g_3 2.012 A_3 $90.8 \cdot 10^{-4}$ cm^{-1}

Magn. Susc.: 98–295 K, p.c. *(8)*
χ_m 1900 · 10^{-6} cm^3 mol^{-1} (295 K)
92–313 K, p.c. *(25)*
χ_m 2052 · 10^{-6} cm^3 mol^{-1} (296.6 K) *(56)*
100–300 K, s.c.
K_x 2702 · 10^{-6} cm^3 mol^{-1} (320 K)
K_y 2338 · 10^{-6} cm^3 mol^{-1} (320 K)
K_z 1797 · 10^{-6} cm^3 mol^{-1} (320 K)

Absorption spectrum: refl., 300 K, 4–14 kK *(7)*
5.60, 8.70, 13.30 kK
refl., 77 K, 3.5–25 kK *(37)*
6.00, 9.00, 13.50 kK

Co(salen)dmso

EPR: dmso, 77 K, g. *(59)*
g_1 2.50 $A_1 < 62 \cdot 10^{-4}$ cm^{-1}
g_2 2.30 $A_2 < 30 \cdot 10^{-4}$ cm^{-1}
g_3 2.013 A_3 $119 \cdot 10^{-4}$ cm^{-1}

Co(salen)CO — EPR: CH_2Cl_2, 130 K, g. (73)

g_1 2.28	A_1 14 $\cdot 10^{-6}$ cm^{-1}	
g_2 2.17	A_2 35 $\cdot 10^{-6}$ cm^{-1}	
g_3 2.02	A_3 80.2 $\cdot 10^{-6}$ cm^{-1}	

Co(salen)CH_3NC — EPR: CH_2Cl_2, 77 K, g. (73)

g_1 2.30	A_1 21 $\cdot 10^{-4}$ cm^{-1}	
g_2 2.19	A_2 34 $\cdot 10^{-4}$ cm^{-1}	
g_3 2.02	A_3 75.4 $\cdot 10^{-4}$ cm^{-1}	

Co(salen)$P(OCH_3)_3$ — EPR: CH_2Cl_2, 77 K, g. (73)

g_1 2.31	A_1 –	A_1^P 234 $\cdot 10^{-4}$ cm^{-1}
g_2 2.18	A_2 33 $\cdot 10^{-4}$ cm^{-1}	A_2^P 253 $\cdot 10^{-4}$ cm^{-1}
g_3 2.02	A_3 72.6 $\cdot 10^{-4}$ cm^{-1}	A_3^P 267 $\cdot 10^{-4}$ cm^{-1}

Co(salen)$P(C_6H_5)_3$ — EPR: CH_2Cl_2, 77 K, g. (73)

g_1 2.40	A_1 –	A_1^P 110 $\cdot 10^{-4}$ cm^{-1}
g_2 2.20	A_2 25 $\cdot 10^{-4}$ cm^{-1}	A_2^P 163 $\cdot 10^{-4}$ cm^{-1}
g_3 2.02	A_3 81.8 $\cdot 10^{-4}$ cm^{-1}	A_3^P 177 $\cdot 10^{-4}$ cm^{-1}

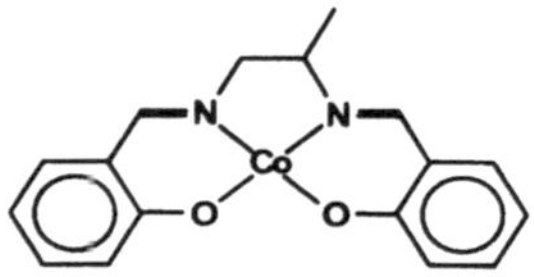

Co(salpn)

Absorption spectrum: $CHCl_3$, 7–36 kK (36)
8.25 (16.8), 27.5 (11500), 28.6 (12600) kK (ϵ)

C.D.: Co(sal(–)pn), $CHCl_3$, 7–36 kK (36)
8.60 (+), 8.86 (– 0.045), 9.30 (– 0.035), 15.50 (0.06), 17.50 (+), 19.00 (+), 26.60 (1.5), 28.26 (1.7) kK ($\Delta\epsilon$)

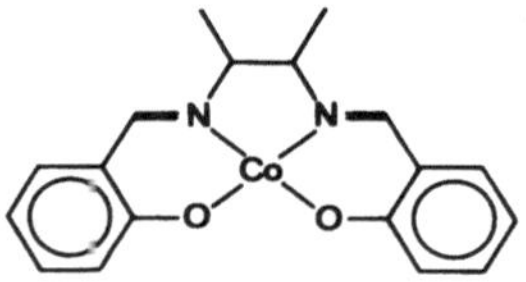

Co(salbn)

ESR: Ni(sal(meso)bn), 100 K, p.c. (7)

g_1 3.37	A_1 179.4 $\cdot 10^{-4}$ cm^{-1}
g_2 1.97	A_2 35.4 $\cdot 10^{-4}$ cm^{-1}
g_3 1.90	A_3 38.7 $\cdot 10^{-4}$ cm^{-1}

C.D.: Co(sal(–)bn), $CHCl_3$, 15–35 kK (7)
18.0 (+ 1.0), 20.3 (3.25), 21.7 (+ 0.85), 22.4 (+ 0.85) kK ($\Delta\epsilon$)

Co(salbn)py — C.D.: Co(sal(–)bn)py, py, 15–35 kK (7)
18.0 (+ 1.40), 19.5 (0.01), 21.0 (– 1.00), 23.0 (+ 0.72) kK ($\Delta\epsilon$)

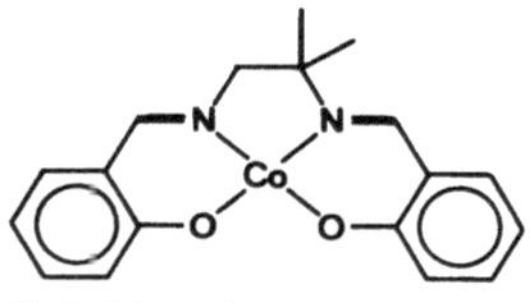

Co(saldmen)

ESR: CH_2Cl_2, 77 K, g. (15)

g_1 3.64	A_1 233 $\cdot 10^{-4}$ cm^{-1}
g_2 1.89	A_2 48 $\cdot 10^{-4}$ cm^{-1}
g_3 1.85	A_3 48 $\cdot 10^{-4}$ cm^{-1}

NMR: $CDCl_3$, 220–320 K (53)

Co(saldmen)py

ESR: 90% thf 10% py, 100 K, g. (*54*)

g_1 2.49
g_2 2.25
g_3 2.01

NMR: py–d_5, 210–350 K (*54*)

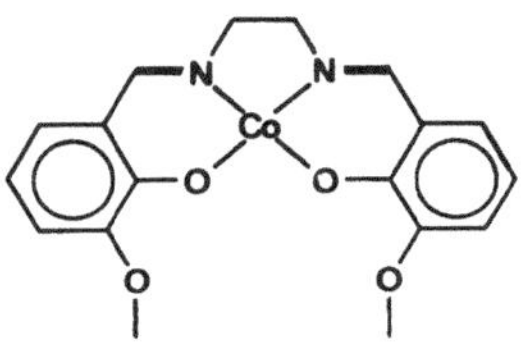

Co(3-methoxysalen)

ESR: Ni(3-methoxysalen)A, 110 K, p.c. (*66*)

g_1	3.430 (5)	A_1	185 (1) · 10^{-4} cm^{-1}
g_2	1.93 (1)	A_2	33 (2)
g_3	1.85 (1)	A_3	25 (2)

Ni(3-methoxysalen)B, 110 K, p.c. (*66*)

g_1	3.247 (5)	A_1	137 (1) · 10^{-4} cm^{-1}
g_2	1.98 (1)	A_2	39 (4) · 10^{-4} cm^{-1}
g_3	1.90 (1)	A_3	27 (1) · 10^{-4} cm^{-1}

Magn. Susc.: 100–296 K, p.c. (*8*)

χ_m 2500 · 10^{-6} cm^3 mol^{-1} (296 K)

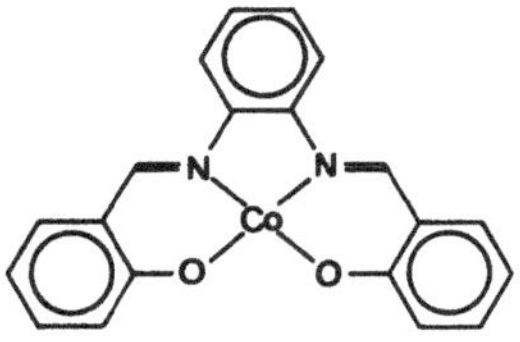

Co(saphen)

EPR: Ni(saphen), 110 K, p.c. (*32*)

g_1	3.845 (5)	A_1	290.2 (5) · 10^{-4} cm^{-1}
g_2	1.75 (2)	A_2	24 (5) · 10^{-4} cm^{-1}
g_3	1.73 (2)	A_3	40 (10) · 10^{-4} cm^{-1}

Absorption spectrum: $CHCl_3$, 3.5–15 kK (*66*)

4.20 (45), 4.80 (12), 8.00 (35) kK (ϵ)

P.E.: (*4*)

Co 2 $p_{1/2}$	796.1 (2), 794.9 (5) eV
Co 2 $p_{3/2}$	779.8 (2) eV
Co 2 $p_{1/2,3/2}$	60.3 (2) eV
N ls$_{1/2}$	398.9 (3) eV
O ls$_{1/2}$	531.2 (1) eV

Co(saphen)py

EPR: 60% toluene 20% CH_2Cl_2 20% py, 130 K, g. (*32*)

g_1	2.407 (15)	A_1	32 (8) · 10^{-4} cm^{-1}
g_2	2.226 (12)	A_2	10 (3) · 10^{-4} cm^{-1}
g_3	2.012 (2)	A_3	91 (2) · 10^{-4} cm^{-1}
		A^N	16.5 (2) · 10^{-4} cm^{-1}

Co(saphen)py_2

EPR: 80% py 20% CH_2Cl_2, 130 K, g. (*32*)

g_1	2.305 (5)	A_1	20 (5) · 10^{-4} cm^{-1}
g_2	2.191 (5)	A_2	38 (5) · 10^{-4} cm^{-1}
g_3	2.035 (2)	A_3	74 (2) · 10^{-4} cm^{-1}
		A^N	11.7 · 10^{-4} cm^{-1}

Co(saphen)$P(OC_2H_5)_3$ EPR: 60% toluene 20% CH_2Cl_2 20% $P(OC_2H_5)_3$, 130 K, g. *(32)*

g_1	2.272 (6)	A_1	$15\ (5) \cdot 10^{-4}\ cm^{-1}$
		A_1^P	$234\ (20) \cdot 10^{-4}\ cm^{-1}$
g_2	2.169 (5)	A_2	$35\ (6) \cdot 10^{-4}\ cm^{-1}$
		A_2^P	$236\ (20) \cdot 10^{-4}\ cm^{-1}$
g_3	2.017 (2)	A_3	$71\ (2) \cdot 10^{-4}\ cm^{-1}$
		A_3^P	$279\ (3) \cdot 10^{-4}\ cm^{-1}$

Co(saphen)thf EPR: 60% toluene 40% thf, 130 K, g. *(32)*

g_1	2.514 (30)	A_1	$58\ (15) \cdot 10^{-4}\ cm^{-1}$
g_2	2.256 (20)	A_2	32.5 (8)
g_3	2.017 (3)	A_3	112 (3)

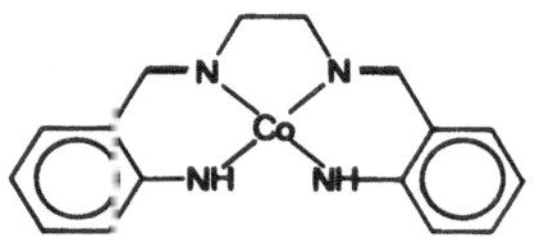

Co(amben)

EPR: Ni(amben), 110 K, s.c. *(47)*

g_x	2.6586 (6)	A_x	$3.5\ (2) \cdot 10^{-4}\ cm^{-1}$
g_y	1.9814 (7)	A_y	$29.5\ (1) \cdot 10^{-4}\ cm^{-1}$
g_z	2.0068 (10)	A_z	$24.0\ (1) \cdot 10^{-4}\ cm^{-1}$

Ni(amben), 110 K, s.c. *(76)*

g_x	2.647 (1)	A_x	$3.5\ (1) \cdot 10^{-4}\ cm^{-1}$
g_y	1.976 (1)	A_y	$31\ (3) \cdot 10^{-4}\ cm^{-1}$
g_z	2.007 (1)	A_z	$21\ (3) \cdot 10^{-4}\ cm^{-1}$

PVC film, 110 K, 110 K *(26)*

g_1	2.65	A_1	$< 10 \cdot 10^{-4}\ cm^{-1}$
g_2	2.03	A_2	–
g_3	1.974	A_3	$28 \cdot 10^{-4}\ cm^{-1}$

Phase 7 A[f]), 110 K, n.p. *(76)*

g_x	2.73 (1)	A_x	$17\ (4) \cdot 10^{-4}\ cm^{-1}$
g_y	1.972 (1)	A_y	$28\ (1) \cdot 10^{-4}\ cm^{-1}$
g_z	1.991 (3)	A_z	$21\ (1) \cdot 10^{-4}\ cm^{-1}$

f) Nematic phases licristal, *Merck*, Darmstadt

P.E.: *(4)*

Co 2 $p_{1/2}$	796.2 (3), 794.2 (2) eV
Co 2 $p_{3/2}$	779.3 (1) eV
Co 2 $p_{1/2,3/2}$	60.0 (1) eV
N $ls_{1/2}$	398.2 (1) eV

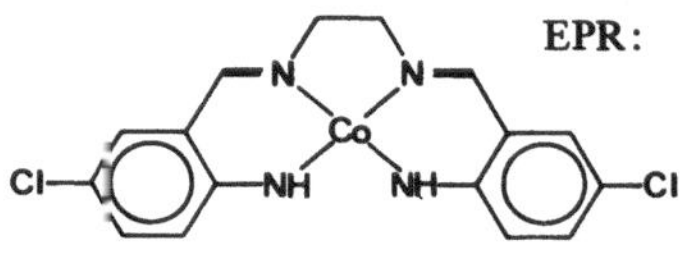

Co(5-Clamben)

EPR: Ni(5-Clamben), 130 K, p. *(26)*

g_1	2.67	A_1	$< 10 \cdot 10^{-4}\ cm^{-1}$
g_2	2.01	A_2	$- \cdot 10^{-4}\ cm^{-1}$
g_3	1.975	A_3	$28 \cdot 10^{-4}\ cm^{-1}$

Magn. Susc.: 300 K, s.c. *(55)*

K_x 1450 · 10^{-6} cm^3 mol^{-1}
K_y 2326 · 10^{-6} cm^3 mol^{-1}
K_z 1437 · 10^{-6} cm^3 mol^{-1}

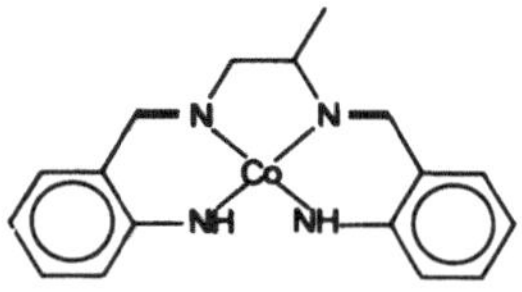

Co(ambpn)

EPR: toluene, 100 K, g. *(70)*

g_x	2.6985	A_x	20 Gauss
g_y	1.9764	A_y	30.9 Gauss
g_z	2.0703	A_z	–

Absorption spectrum: C_2Cl_4, 4–30 kK *(70)*
5.10 (18), 6.25 (22), 6.90 sh, 11.05 (73), 11.90 sh (48), 16.67 sh (370), 18.90 (500), 23.15 (22000), 26.90 (15500) kK (ϵ)

C.D.: Co(amb(–)pn), C_2Cl_4, 8–35 kK *(70)*
10.74 (0.62), 11.36 (– 0.09), 15.43 sh (0.24), 16.89 (1.62), 20.00 (3.75), 21.69 (3.25), 23.26 (3.39), 25.97 (5.19), 30.30 (7.20) kK ($\Delta\epsilon$)

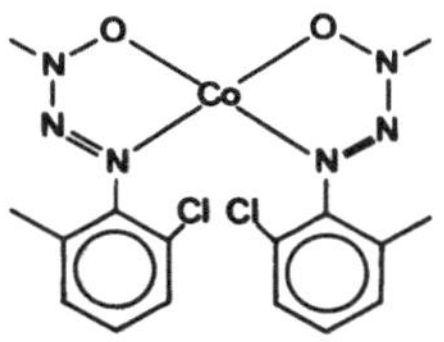

$Co(pto)_2$

EPR: $Ni(pto)_2$, 110 K, s.c. *(14)*

g_x	2.806	A_x	16.5 · 10^{-4} cm^{-1}
g_y	1.966	A_y	36.0 · 10^{-4} cm^{-1}
g_z	2.062	A_z	26.6 · 10^{-4} cm^{-1}

$Ni(pto)_2$, 110 K, p.c. *(14)*

g_1	2.654	A_1	17.0 · 10^{-4} cm^{-1}
g_2	2.081	A_2	26.4 · 10^{-4} cm^{-1}
g_3	1.980	A_3	35.9 · 10^{-4} cm^{-1}

Absorption spectrum: $CHCl_3$, 4–35 kK *(14)*
4.65 (19), 11.76 (12), 16.10 (44), 19.40 (350), 23.25 (820), 30.30 (5800) kK (ϵ)

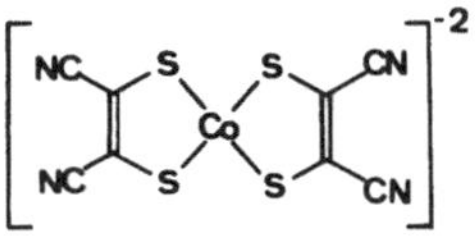

$[Co(mnt)_2]^{2-}$

EPR: $Ni(mnt)_2^{2-}$, 110 K, s.c. *(46)*

g_x	2.798 (3)	A_x	50 (1) · 10^{-4} cm^{-1}
g_y	1.977 (3)	A_y	28 (1) · 10^{-4} cm^{-1}
g_z	2.025 (3)	A_z	23 (1) · 10^{-4} cm^{-1}

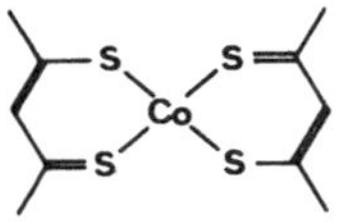

$Co(sacsac)_2$

EPR: $Ni(sacsac)_2$, 130 K, s.c. *(30)*

g_x	3.280	A_x	105 · 10^{-4} cm^{-1}
g_y	1.904	A_y	35 · 10^{-4} cm^{-1}
g_z	1.899	A_z	35 · 10^{-4} cm^{-1}

References

1. *Amos, A. T., Hall, G. G.:* Proc. Roy. Soc., Ser. A, *263*, 483 (1961)
2. *Assour, J. M., Goldmacher, J., Harrison, S. E.:* J. Chem. Phys. *43*, 159 (1965)
3. *Basch, H., Viste, A., Gray, H. B.:* J. Chem. Phys. *44*, 10 (1966)
4. *Burness, J. H., Dillard, J. G., Taylor, L. T.:* Inorg. Nucl. Chem. Lett. *9*, 825 (1973)
5. *Bürgi, H. B.:* Swiss Federal Institute of Technology, Zürich, personal communication
6. *Busetto, C., Cariati, F., Fantucci, P. C., Galizzioli, D., Morazzoni, F., Valenti, V.:* Gazz. Chim. Ital. *102*, 1040 (1972)
7. a) *Busetto, C., Cariati, F., Fantucci, P. C., Galizzioli, D., Morazzoni, F.:* Inorg. Nucl. Chem. Lett. *9*, 313 (1973)
   b) *Busetto, C., Cariati, F., Fantucci, P. C., Galizzioli, D., Morazzoni, F.:* J. Chem. Soc. Dalton: 1712 (1973)
   c) *Busetto, C., Cariati, F., Fusi, A., Gullotti, M., Morazzoni, F., Pasini, A., Ugo, R., Valenti, V.:* J. Chem. Soc. Dalton: 754 (1973)
8. a) *Calvin, M., Bailes, R. H., Wilmarth, W. K.:* J. Am. Chem. Soc. *68*, 2254 (1946)
   b) *Calvin, M., Barkelew, C. H.:* J. Am. Chem. Soc. *68*, 2267 (1946)
9. *Cariati, F., Sgamelotti, A., Valenti, V.:* Atti. Accad. Naz. Lincei, Rend., Cl. Sci. Fis. Nat. Natur. *45*, 344 (1968)
10. *Carter, M. J., Rillema, D. P., Basolo, F.:* J. Am. Chem. Soc. *96*, 392 (1974)
11. *Carlisle, G. O., Simpson, G. D., Hatfield, W. E., Crawford, V. H., Drake, R. F.:* Inorg. Chem. *14*, 217 (1975)
12. *Cariati, F., Morazzoni, F., Busetto, C., Del Piero, G., Zazzetta, A.:* J. Chem. Soc. Dalton: 342 (1976)
13. *Calligaris, M., Minichelli, D., Nardin, G., Randaccio, L.:* J. Chem. Soc. A. 2411 (1970)
14. *Chacko, V. P., Manoharan, P. T.:* J. Magn. Reson. *16*, 75 (1974)
15. *Chikira, M., Kawakita, T., Isobe, T.:* Bull. Chem. Soc. Jpn., *47*, 1283 (1974)
16. *Chikira, M., Migita, K., Kawakita, T., Iwaizumi, M., Isobe, T.:* J. Chem. Soc. Chem. Commun. 316 (1976)
17. *Ciampolini, M.:* Structure and Bonding *6*, 52 (1969)
18. *Clark, G. R., Hall, D., Waters, T. N.:* J. Chem. Soc. A. 223. (1968)
19. *Clementi, E., Raimondi, D. L.:* J. Chem. Phys. *38*, 2686 (1963)
20. *Cotton, F. A., Wilkinson, G.:* Advanced Inorganic Chemistry, New York–London–Sydney–Toronto: J. Wiley & Sons 1972
21. *Daul, C.:* Dissertation Nr. 724, University of Fribourg, Switzerland (1974)
22. *Daul, C.:* unpublished results.
23. *Dedieu, A., Rohmer, M. M., Veillard, A.:* J. Am. Chem. Soc. *98*, 5789 (1976)
24. *DeIasi, R., Holt, S. L., Post, B.:* Inorg. Chem. *10*, 1498 (1971)
25. *Earnshaw, A., Hewlett, P. C., King, E. A., Larkworthy, L. F.:* J. Chem. Soc. A 241. (1968)
26. *Engelhardt, L. M., Duncan, J. D., Green, M.:* Inorg. Nucl. Chem. Lett. *8*, 725 (1972)
27. *Fantucci, P. C., Valenti, V., Cariati, F.:* Inorg. Nucl. Chem. Lett. *11*, 585 (1975)
28. *Fantucci, P. C., Valenti, V.:* J. Am. Chem. Soc. *98*, 3832 (1976)
29. *Fierz, H.:* Dissertation Nr. 699, University of Fribourg, Switzerland (1972)
30. *Gregson, A. K., Martin, R. L., Mitra, S.:* Chem. Phys. Lett. *5*, 310 (1970)
31. *Griffith, J. S.:* The Theory of Transition-Metal Ions. Cambridge: University Press 1961
32. *Haas, O.:* Dissertation Nr. 753, University of Fribourg, Switzerland (1976)
33. *Hall, D., Rae, A. D., Waters, T. N.:* J. Chem. Soc. 5897. (1963)
34. *Hall, D., Morgan, H. J., Waters, T. N.:* J. Chem. Soc. A 677. (1966)
35. *Harnung, S. E., Schäffer, C. E.:* Structure and Bonding *12*, 257 (1972)
36. *Hipp, C. J., Baker, W. A., Jr.:* J. Am. Chem. Soc. *92*, 792 (1970)
37. *Hitchman, M. A.:* Inorg. Chem. *16*, 1985 (1977)
38. *Hitchman; M. A.:* Inorg. Chim. Acta, *26*, 237 (1978)

39. *Hoffmann, R.:* J. Chem. Phys. *39*, 1397 (1963)
40. *Hoffman, B.M., Basolo, F., Diemente, D.L.:* J. Am. Chem. Soc. *95*, 6497 (1973)
41. *Holm, R.H., Everett, G.W., Jr., Chakravorty, A.:* Progr. Inorg. Chem. *7*, 83 (1966)
42. *Jϕrgensen, C.K.:* Inorganic complexes, London–New York: Academic Press 1963
43. *Jϕrgensen, C.K.:* Modern Aspects of Ligand Field Theory. Amsterdam: North Holland Publishing Company 1971
44. *Jungo, Ch.:* Dissertation Nr. 775, University of Fribourg, Switzerland (1977)
45. *Kataoka, N., Kan, H.:* J. Am. Chem. Soc. *90*, 2978 (1968)
46. *Maki, A.H., Edelstein, N., Davison, A., Holm, R.H.:* J. Am. Chem. Soc. *86*, 4580 (1964)
47. *Malatesta, V., McGarvey, B.R.:* Can. J. Chem. *53*, 3791 (1975)
48. *Marzilli, L.G., Marzilli, P.A.:* Inorg. Chem. *11*, 457 (1972)
49. *McGarvey, B.R.:* Electron Spin Resonance of Transition-Metal Complexes. In Transition Metal Chemistry, Vol. 3 (Carlin, R.L., ed.). pp. 90–194. New York: Dekker 1966
50. *McGarvey, B.R.:* Can. J. Chem. *53*, 2498 (1975)
51. *Montgomery, H., Morosin, B.:* Acta Crystallogr. *14*, 551 (1961)
52. *McHugh, A.J., Gouterman, M., Weiss, Ch., Jr.:* Theor. Chim. Acta, *24*, 346 (1972)
53. *Migita, K., Iwaizumi, M., Isobe, T.:* J. Am. Chem. Soc. *97*, 4228 (1975)
54. *Migita, K., Iwaizumi, M., Isobe, T.:* J. Chem. Soc. Dalton 532 (1977)
55. *Murray, K.S., Sheahan, R.M.:* J. Chem. Soc. Chem. Commun. 475 (1975)
56. *Murray, K.S., Sheahan, R.M.:* J. Chem. Soc. Dalton 999, 1134 (1976)
57. *Nishikawa, H., Yamada, S.:* Bull. Chem. Soc. Jpn. *37*, 8 (1964)
58. *Nishida, Y., Kida, S.:* Bull. Chem. Soc. Jpn. *45*, 461 (1972)
59. *Ochiai, E.:* J. Inorg. Nucl. Chem. *35*, 1727 (1973)
60. *Otsuka, J.:* J. Phys. Soc. Jpn. *21*, 596 (1966)
61. *Perumareddi, J.R.:* Coord. Chem. Rev. *4*, 73 (1969)
62. *Pfeiffer, P.E., Breith, E., Lübbe, E., Tsumaki, T.:* Justus Liebigs Ann. Chem. *503*, 84 (1933)
63. *Pople, J.A., Beveridge, D.L., Dobosh, P.A.:* J. Chem. Phys. *47*, 2026 (1967)
64. *Richardson, J.W., Nieuwpoort, W.C., Powel, R.R., Edgell, W.F.:* J. Chem. Phys. *36*, 1057 (1962)
65. *Schaeffer, W.P., Marsh, R.E.:* Acta Cristallogr. Sect. B, *25*, 1675 (1969)
66. *Schläpfer, C.W.:* to be published
67. *Scullane, M.I., Allen, H.C., Jr.:* J. Coord. Chem. *4*, 255 (1975)
68. *Srivanavit, C., Brown, D.G.:* Inorg. Chem. *14*, 2950 (1975)
69. *Srivanavit, C., Brown, D.G.:* J. Am. Chem. Soc. *98*, 4447 (1976)
70. *Urbach, F.L., Bereman, R.D., Topich, J.A., Hariharan, M., Kalbacher, B.J.:* J. Am. Chem. Soc. *96*, 5063 (1974)
71. *Viste, A., Gray, H.B.:* Inorg. Chem. *3*, 1113 (1964)
72. *Walker, F.A.:* J. Am. Chem. Soc. *92*, 4235 (1970)
73. *Wayland, B.B., Abd-Elmageed, M.E., Mehne, L.F.:* Inorg. Chem. *14*, 1456 (1975)
74. *von Zelewsky, A.:* Helv. Chim. Acta, *55*, 2941 (1972)
75. *von Zelewsky, A., Fierz, H.:* Helv. Chim. Acta, *56*, 977 (1973)
76. *Zobrist, M.:* Dissertation Nr. 732, University of Fribourg, Switzerland (1974)

Author-Index Volumes 1 — 36